W9-CXN-937

OIL AND POLITICS

Oil and Politics

HENRI MADELIN

Translated from the French by
MARGARET TOTMAN

SAXON HOUSE | LEXINGTON BOOKS

First published as Pétrole et Politique en Mediterranée Occidentale by Armand Colin

Published by

SAXON HOUSE, D. C. Heath Ltd.
Westmead, Farnborough, Hants., England

Jointly with

LEXINGTON BOOKS, D. C. Heath & Co.
Lexington, Mass., U.S.A.

ISBN 0 347 01077 6
Library of Congress Catalog Card Number 74-31975

Printed in Great Britain
by Unwin Brothers Limited
The Gresham Press, Old Woking, Surrey
A member of the Staples Printing Group

Contents

List of figures

List of tables

Introduction: The Participants in Oil Affairs

> Oil occupies a unique place in the popular imagination: it is always topical. It is worshipped and it is feared. There is almost fanatical glorification of its potentially beneficial forces; conversely there is extreme exaggeration of its power to cause harm. People readily imagine that where oil is concerned anything may happen . . . and, indeed, it does.
>
> P.H. Frankel, *Essentials of Petroleum, a Key to Oil Economics,* Chapman & Hall, London 1946, p. 11

The developed areas of the world have suffered violent repercussions since the third Arab-Israeli war, in October 1973. After Arab producer countries had threatened to use oil as a political weapon and had imposed a temporary embargo, the price of oil was quadrupled over a period of a few months. Some of the consequences of this upheaval in the oil market are already apparent in 1974; other effects not immediately foreseen will follow inescapably in the course of time. The old fear of energy shortage in industrialised civilisations seemed to have been banished but has returned, now taking the form of a threat to their security. Oil is front page news and has become a favourite topic for expert studies and scientific discussion.

This book is a further contribution to the vast quantities of material on the subject now being assembled as everyone seeks to analyse the reasons for the current situation and to envisage future developments.

In order to understand the present situation in connection with oil it is necessary to study the past. The changes now taking place can be appreciated only by reference to the history of the men, the organisations and the policies which have shaped world oil affairs since the beginning of this century.

The voracious appetite for energy in industrialised societies and the growing nationalism of the producing countries in energy matters have caused oil to become a prominent political issue. My aim in this book is to assess the true place of oil in the world. Specialists tend to limit their field of study to technical, economic or geographical considerations. This account seeks to sweep away such confines and provide a political perspective that has hitherto largely been indiscernible in books on the subject. In considering why a public sector in the form of state companies has recently come to the fore in oil affairs and in seeking to understand the reasons for increasing intervention by governments it is necessary to adopt a broad general view of the oil phenomenon. As an energy source of growing importance in economic life, oil takes on a political aspect, at any rate in politics in the widest meaning of the term, for around oil as a raw material there arise relations of co-operation and of conflict between men, organisations and states.

Some of the traditional participants are involved wherever oil politics occur but for many years now a newcomer, the state company, with a special rôle as an instrument of national policies, has been trying to join in. The political accompaniment to economic affairs has been accentuated with the emergence of the state companies and is especially apparent in the oil affairs of the countries either side of the Western Mediterranean.

The participants in any oil politics

It is impossible to discuss oil without mentioning the strategies of the groupings concerned. Oil politics result from a subtle kind of triangular interplay between three main contenders: the big international concerns, whose economic activities necessarily have political consequences; the producing countries, having political ambitions to a greater or lesser degree compatible with the constraints of the national economy; and the consumer countries, where the State has the power to impose economic and political rationality, even if, in practice, it shows greater concern for acting in the general interests of the population as a whole.

One can see at a glance that in basic issues of prices and results the interests of each of the three protagonists are not spontaneously harmonious. The producing countries are above all concerned with taxation, constantly seeking to improve their share of profit from the companies' oil operations. For their part, the big companies aim to maximise the quantities sold, in order to minimise the unit cost. However, what the consumers need most is a combination of security of supplies with the lowest possible price.

The large companies

For a long time the world oil market has been dominated by the big capitalist companies operating on an international scale and usually referred to generically as the majors. These concerns form self-sufficient decision making centres with worldwide strategy.

Concerning the way in which the power of the big private groups has evolved from the history of the oil industry, nobody has expressed the matter more clearly than Enrico Mattei, since the whole policy of his organisation may be summed up as an attempt to challenge the situation which existed at the end of the Second World War. In a speech given in 1959 he said:

> That the oil industry has become increasingly concentrated in the hands of a small group of vertically integrated organisations which are banded together in various ways and that for a long time satisfaction of the oil needs of the whole world has almost entirely depended on these, and still does to a large extent to-day, is due to the conditions in which the oil industry began and grew up.
>
> Undoubtedly the system met military needs and those of a world economy in the developing stage. But it is equally undeniable that consumers of petroleum products have paid dearly for the services rendered.
>
> It is worth noting that for the very small number of large international companies – most of them also engaged in the oil industry in the United States – dominating world crude oil production there was every interest in exploiting the resources available in all areas, without boosting output in any of them in particular. For some thirty years now these companies have, in fact, functioned as regulators of production, sharing coverage of the world demand determined by the level of prices and by the fiscal policy of each of the consumer countries between the different areas and the various producer companies.
>
> Clearly, this state of affairs has not arisen from a concerted plan, but a set of circumstances and a network of tacit agreements have had the outcome at any given moment of eliminating competition between the big companies and potential challenge to them from smaller organisations.
>
> But the system has also been based on interests of the major international oil groups coinciding with the national interests of the United States of America, namely that the selling price of crude should be high enough to make extraction profitable even in continental America.[1]

This castigation would seem to be justified at any rate in relation to the period before 1956–60. The points made give an indication of how the setting up of a state company is linked with a desire for national independence.

Yet one factor omitted from this overall view is the financial power of the big groups, which is both the cause and the result of the scope of their policies. To illustrate this aspect it may suffice to quote the figures given by Maurice Byé in 1960:

> Standard Oil of New Jersey, with annual gross receipts of nearly 6 thousand million dollars, is comparable in finance to a nation like Canada. It represents one-fifth of world oil affairs, controls sources of production in five countries and means of transport, refining and distribution in most of the other countries.[2]

Going on from this one company to a wider view, one finds that in 1949 the seven leading oil concerns of the world, collectively called 'the majors' by experts, held 34 per cent of reserves and 32 per cent of production within the United States. On the American market they were not entirely dominant because despite their extreme concentration they were in competition with 12,500 domestic producers.[3] But their position outside the United States is different.

The addition to the list of the seven largest organisations of an eighth, namely Compagnie Française des Pétroles, gives an even clearer picture of the power and concentration of the big enterprises. In 1965 and 1966 the eight leading international companies were responsible for about 63 per cent of world production, 60 per cent of refining and more than 62 per cent of distribution (excluding the Communist bloc).[4] In 1968 the seven majors accounted for 60 per cent of world production, 56·2 per cent of refining and 55·4 per cent of distribution outside the Communist bloc. The same companies were well established in the most productive areas of the globe, accounting for 86 per cent of production in the Middle East, 58 per cent in Libya and 90 per cent in Venezuela.

Hence, where the state companies, as comparative newcomers, seek to compete on equal terms with the giants of the oil world there is a certain minimum size below which their influence and efficiency is too limited.

Moreover, as Enrico Mattei explained clearly elsewhere in the speech quoted above, the big international companies have the entirely rational ambition of covering their marginal cost. For that reason they strive in the same way as, for example, the large-scale French agricultural organisations to secure the profitability of the smallest unit at the least productive site, in order to build up profits with which to finance exploration that in any circumstances is extremely costly. They can also benefit from the experience on which one of the principles of American oil policy is based,

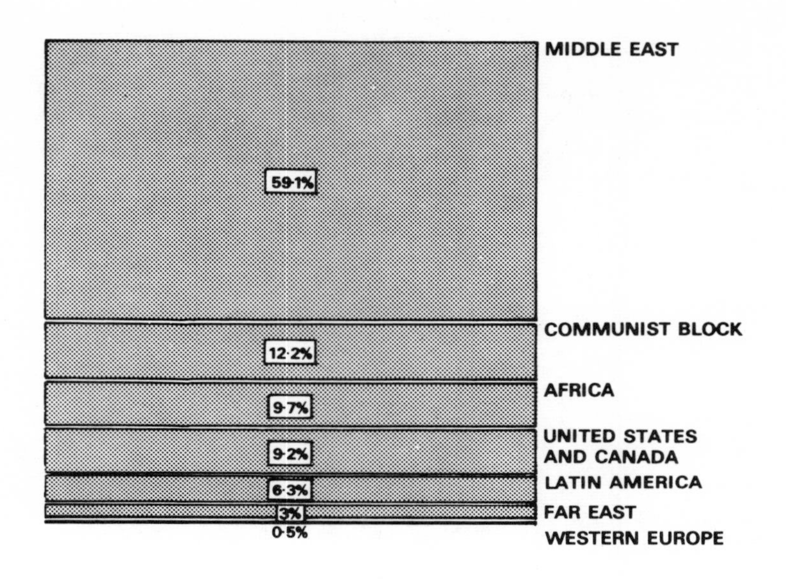

Fig. 1(a) Geographical distribution of world oil reserves

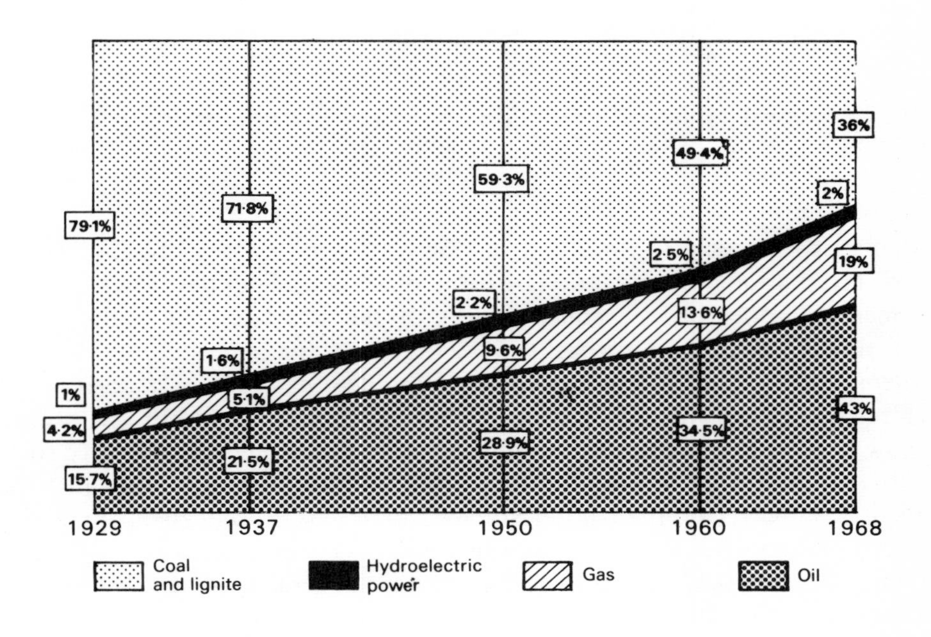

Fig. 1(b) Consumption of primary energy – pattern 1929–68

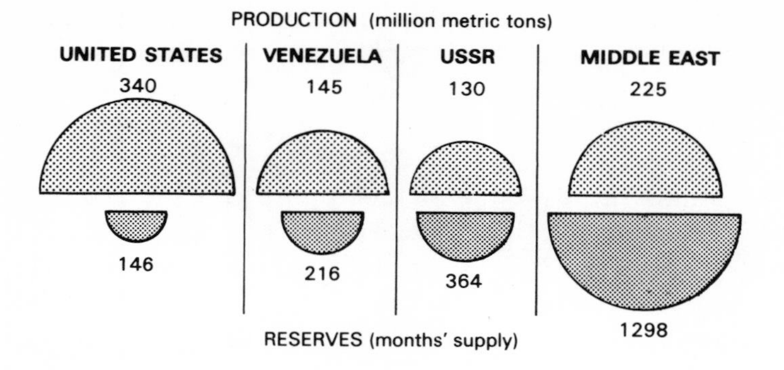

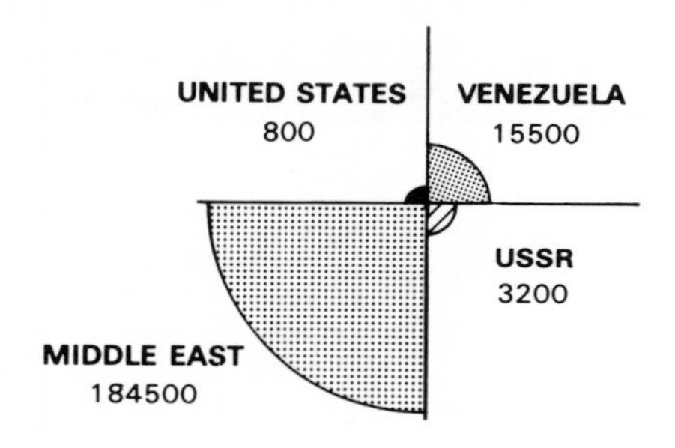

Fig. 2 World crude oil production: reserves and production per well in 1959

that it is necessary to avoid exhausting reservoirs too quickly. Because of the international ramifications of oil operations and the power of the American concerns, the consequences of the situation can be disastrous for non-American enterprises: the big oil-producing organisations, most of them with parent companies in the United States, are in a position to impose their will on all hydrocarbon consumers throughout the world. Now the differences in costs and productivity of wells varies greatly from one continent to another. Estimates of the daily output of wells in 1950 were 11·6 barrels in the United States, 200 barrels in Venezuela and about 5,000 in the Middle East. According to the conclusions of the Federal Trade Commission, the production cost per barrel at that time was 30 cents in Arabia, 50 cents in Venezuela and 1·85 dollars in Texas.[5]

Figure 2 illustrates, for 1959, comparisons between the United States and three other oil-producing regions of the world – Venezuela, the USSR and the Middle East – in production and reserves and in productivity per well. The oil problem in the United States is immediately apparent: production is over-intensive by comparison with the mean per well elsewhere and in relation to estimated reserves calculated in monthly units.

By 1969 Middle East production comprised an even larger percentage of the world total but the United States domestic percentage had dropped considerably (see figure 3). According to the experts the relative difference in productivity remained similar to that in 1959, with daily production per well in the order of 2 metric tons in the United States, against 540 in the Middle East. It is judged that 80 per cent of the wells in the United States would be unprofitable in the event of completely free competition.

A report published in 1974 put the current number of producing oil-wells at 503,500 but pointed out that 'about three-quarters of them (some 353,000) are "strippers" with a marginal production averaging only 3·7 barrels a day per well.' The report went on to explain that:

> These 'strippers' account for about 13 per cent of total US production and have the effect of reducing the average daily production of oil per well to 18·8 barrels. Because of the lack of economic incentive to continue stripper production, it has been estimated early in 1973 that about 20,000 a year of these wells might have to be abandoned. But the increasing price of crude oil and the effect of the Arab embargo on supplies to the USA has given a new lease of life to stripper production, with the possibility of opening up some wells that had previously been plugged.[6]

The discoveries in Alaska were therefore welcomed with some sense of relief, since they improved the proportion of known reserves to annual consumption – a proportion which was then causing some concern about the future of operations in United States territory.

The United States was faced with a real dilemma about oil. Should there be a strict control policy such as the mandatory oil import quota system adopted in March 1959 and still in force when reviewed by a special federal committee in 1969? Whilst providing a large measure of security, that system was causing a great loss of efficiency in internal production and above all it was expensive to the American consumer. If quota and import restrictions were removed or relaxed there would be a temporary drop in crude oil prices in the United States, although American observers predicted that eventually adjustments in output by the two major oil-producing states, Texas and Louisiana, would hoist domestic prices even higher than the current level. For the time being, however, foreign-produced oil would have an advantage over home-produced supplies: it would be cheaper by almost a third to ship Middle East oil into the American eastern cities. Yet the federal authorities feared the security risks of making a zone with enormous oil consumption (stretching from Florida to Maine's Canadian border) heavily dependent on supplies highly vulnerable in the event of an international oil crisis. The dilemma which United States politicians hesitated to solve was whether to choose the greater security of costly domestic production or to accept the possible risks attaching to cheaper supplies.[7] Meanwhile they allow the big American oil groups to go on spreading the web of their gigantic interests over the whole world.

The awakening of the producing countries

For a long time the pricing arrangements for oil were largely based on safeguarding the organisational system, the production costs and the productivity of the United States domestic oil industry. This method was justifiable so long as America played by far the largest part in supplying the international markets. But for some fifteen years now this situation has no longer been tenable, partly because of the tremendous growth of demand in the world as a whole, and above all in Europe, and also because of the increased dynamism or entry into the arena of producer countries geographically distant from the United States market and much closer to the big European centres of consumption. The abundance of their oil resources enables these countries to achieve considerably lower production costs and with better productivity.

As long ago as 1960 an article by two leading oil experts, P.H. Frankel and W.L. Newton, pointed out that it was no longer possible to treat the world market for oil as a single market. There were three separate spheres to consider: the United States, the Soviet bloc and the remaining countries of the world. Both the United States and the Soviet bloc were potentially self-sufficient, the former now being a net importer and the latter a net exporter. The characteristic feature of oil markets in the remaining areas of the world was polarisation of supply and demand: the main producers – Venezuela, the Middle East and North Africa – had no sizeable demand for oil, and the biggest consumers, notably Western Europe, had little or no oil production. During the 1950s there had been a shift in the oil industry from the Western to the Eastern hemisphere.[8]

The Middle East has been in the best position to benefit from the transformation of the situation which has taken place since 1950, especially in view of Japan's enormous demand for raw materials. The Middle East is not only favourably placed geographically but also has exceptionally low production costs. In 1956 there were some 555,000 productive wells in the United States and 678 in the Middle East as a whole. Daily production was 1·8 metric tons in the United States and more than 1,700 metric tons in Iraq, for example. Estimates made by the Chase Manhattan Bank at the end of 1955 give a still clearer picture of the comparison: 23,925 million dollars had been invested in crude oil production in the United States (for sinking wells and for costs of production equipment and excluding geological and geophysical exploration) compared with only 950 million dollars in the Middle East. The production being achieved with these investments was 11,400 million metric tons in the United States and 14,000 million metric tons in the Middle East. Even taking into account the differing duration of exploitation of resources in the two regions, these figures show the great disparity in profitability between the two kinds of reservoir.[9]

According to estimates made in 1974 by the French *Comité professionel du pétrole,* proved oil reserves in the Middle East were more than 50 thousand million metric tons, whilst those in the United States were less than 5 thousand million metric tons. Because the Middle East

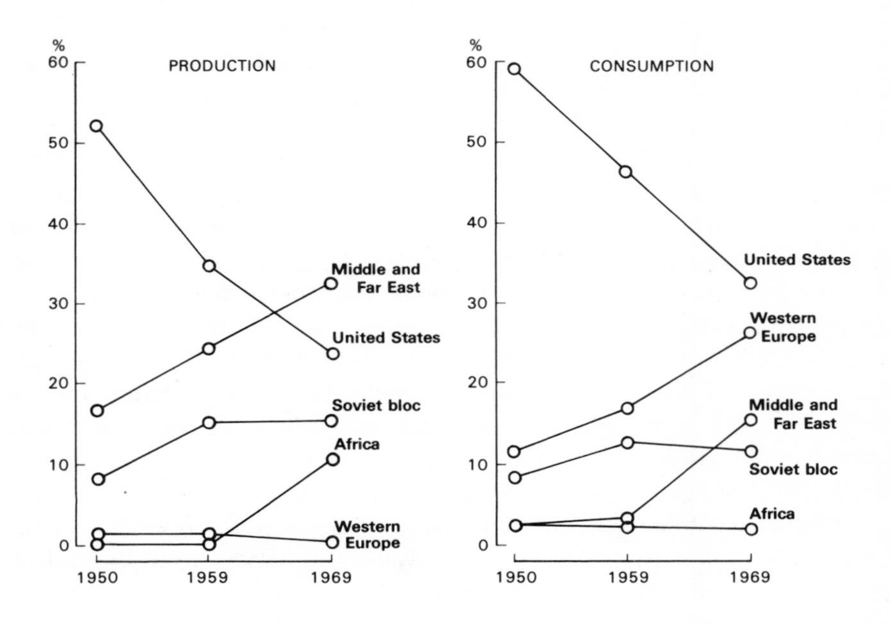

Fig. 3 Oil production and consumption, 1950, 1959 and 1969

countries produce a third of the non-Communist oil and hold 60 per cent of world reserves, this region of the world becomes 'the major centre of current oil geopolitics'.[10]

This irresistible rise of the Middle East has led to substantial changes in the cost structure compared with the previous American hegemony. The price system devolving from the Achnacarry agreement of 1928, when circumstances favoured a cartel, have gradually had to be revised. The agreement between the giants of the oil industry produced the arrangement that the price of oil should be the same in every export centre of the world as in the American ports along the Gulf of Mexico, wherever the oil actually came from. This system, used to peg pricing all over the world, gave the 'Gulf plus' quotation.

In practice, this complicated system meant that the price payable in any country for imported oil was calculated according to a basic price in the Gulf of Mexico. The freight cost added was simply the cost of shipping the oil from the Gulf of Mexico to the point of delivery irrespective of the true point of origin of the oil. As J. Chardonnet comments, 'The net effect was as if all the oil consumed in the world came from the Gulf of Mexico, whatever its true origin'.[11]

This system had the disadvantage of completely disregarding the true costs of production and the actual transportation costs: a current phrase was 'phantom freight'. It did not matter overmuch while the United States dominated production and the export trade in oil. But once the Middle East became a big producer and Europe's consumption surged ahead the defects of this artificial arrangement became obvious. By reason of its proximity to Europe and above all because of its greater profitability, the selling price of oil from the East came out of the American sphere of influence. In 1943 a second basing point was established in the Persian Gulf, following massive purchases of fuel for the British and United States Navies. From that time on the break-even line moved steadily westwards, first to Italy and then to Britain in 1947 and finally in 1949 even reaching the Eastern Seaboard of the United States and the wall of American protectionism.[12] The previous arrangements were swept aside by the rising importance of Middle East oil but a relatively artificial price system was still needed. It still operates to-day in posted prices.

However, the Middle East is now subjected to vigorous competition from relative newcomers to oil production – the North African countries and Nigeria. Oil was first discovered in Algeria in 1955 and in Libya four years later and in 1969 the combined output from these two countries accounted for more than 9 per cent of the world production. Because they were in any case relatively close to the big European markets and,

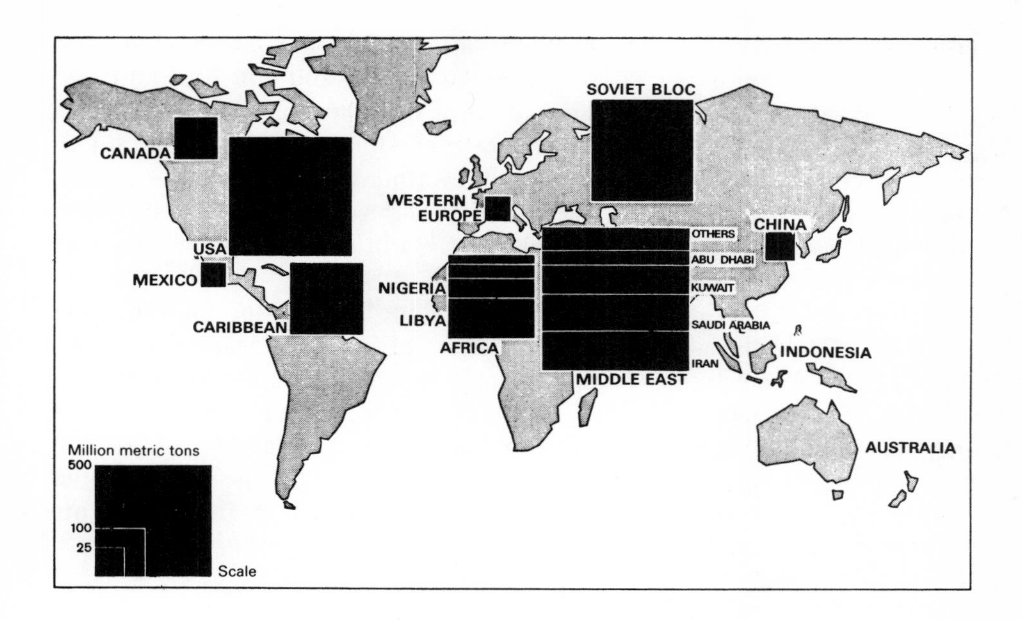

Fig. 4 Crude oil production, 1971

Source: Comité professionnel du pétrole (French oil trade committee)

indeed, were in a fortunate geographical position when the Suez Canal was closed, Algerian and Libyan oil supplies in turn caused an upheaval in the international market. Libya, with the additional advantage of oilfields conveniently near the sea, had become the more prominent producer of the two in 1969, with an output of more than 150 million metric tons.

The change in the supply situation brought about an awakening of the host countries, with increasing aspirations that the riches in the subsoil should help to raise the status of these countries and to give them effective bargaining power at international level. A new 'oil nationalism' began to emerge. Among the first spokesmen for this ideology were ex-President Betancourt of Venezuela [13] and Sheikh Tariki, former Saudi Oil Minister. Sheikh Tariki expressed the view that sovereign states did not need to negotiate with private enterprises, however important, but could unilaterally impose their will on them.[14]

However, the governments of the producer countries have not yet adopted such extreme positions: their attitude is usually more moderate. They have not forgotten the disastrous results of Dr Mossadegh's experiment in nationalisation of Iranian oil in 1951.

For the time being it is difficult to avoid playing along with the big oil groups, which have several trumps that can be used in the event of conflict, including solidarity between the groups and the possibility of taking advantage of the transnational scope of operations which enables a group to step up output in one country and to reduce production in another country which is too demanding. Rather than attempt to oust the big companies the host countries often prefer to put pressure on them with a view to obtaining more than the traditional fifty-fifty share of profits.

Things are not working out altogether smoothly for the oil exporting countries. There are all too many instances of lack of harmony between states on the same side who are scarcely able to agree on common strategy. There seems to be great difficulty in working out arrangements and policies enabling them to achieve flourishing national companies. The problem for these companies is how to gain a place in the international market without causing a drop in the price levels which their governments are striving to maintain come what may.

The consumer countries

For the third main partner in the relationship things have taken a turn for the better since 1960. This is because the consumer countries have growing markets which can use and pay for increasing quantities of crude oil. This is especially true of Europe. In 1968 there was an 11 per cent increase in internal consumption in the Common Market, which was almost double the growth rate for the United States. [15] Hence a shift in the focal point of the oil trade, transferring priority to relations between Europe and the African countries or the Persian Gulf states instead of favouring links between the United States and Europe.

Whilst there may be cost advantage in obtaining supplies from the Middle East and to a lesser degree from the African states, it must be emphasised that the best tactic for the importing countries in this three-sided relationship is to concentrate on consumer protection.

After the end of the last world war there was first of all grumbling and then more open rebellion in the consumer countries against the domination of the big companies. Under the set-up existing at that time consumers either side of the Atlantic were confronted by the transnational policies of the majors. Various reports which received a great deal of publicity took up the cudgels on behalf of European and American consumers. In 1946 a committee investigated oil prices in the Persian Gulf and revealed that a strange set of rates based on the artificial 'Gulf plus' system had been devised for supplies to the British Navy during the war.

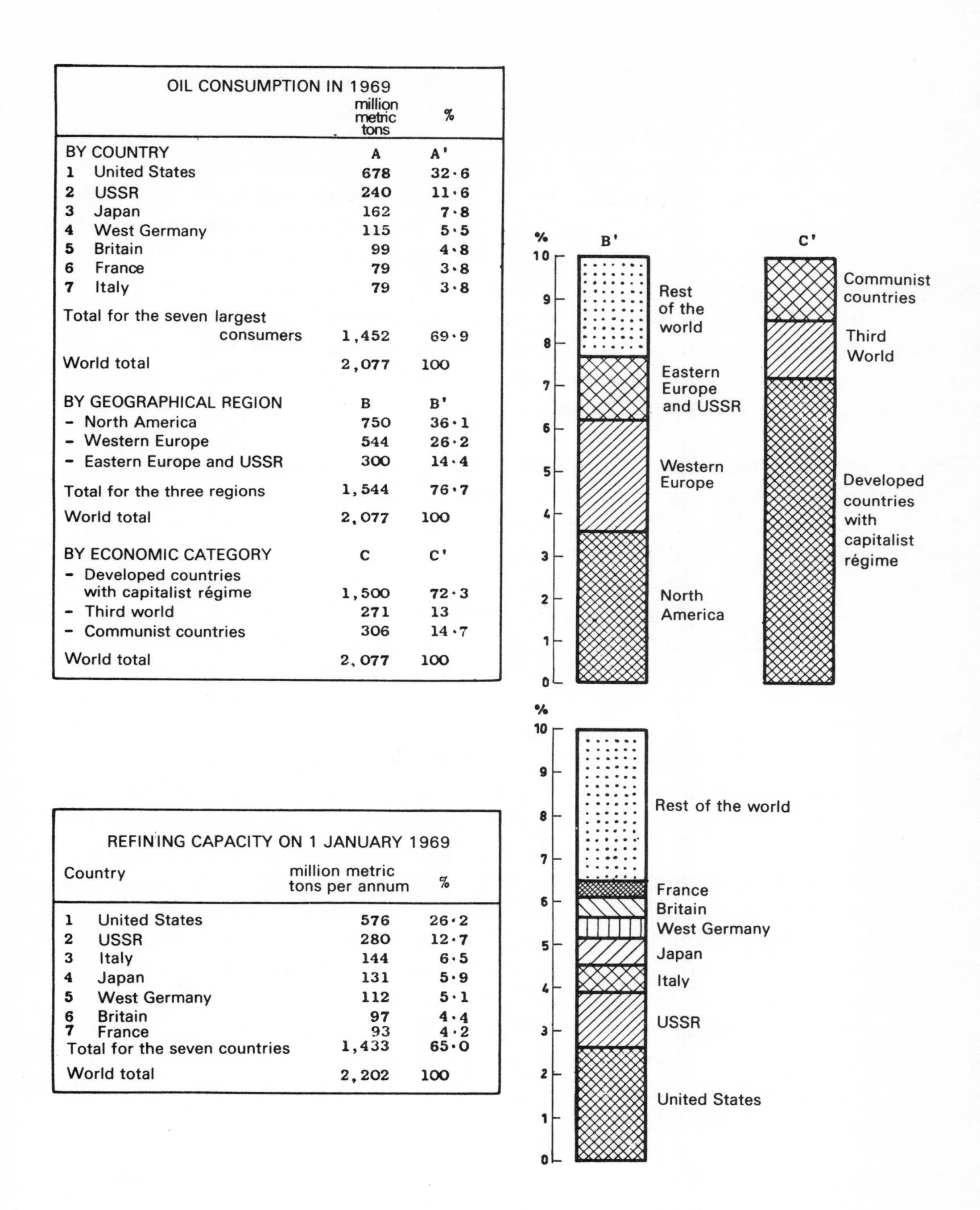

OIL CONSUMPTION IN 1969	million metric tons	%
BY COUNTRY	A	A'
1 United States	678	32·6
2 USSR	240	11·6
3 Japan	162	7·8
4 West Germany	115	5·5
5 Britain	99	4·8
6 France	79	3·8
7 Italy	79	3·8
Total for the seven largest consumers	1,452	69·9
World total	2,077	100
BY GEOGRAPHICAL REGION	B	B'
- North America	750	36·1
- Western Europe	544	26·2
- Eastern Europe and USSR	300	14·4
Total for the three regions	1,544	76·7
World total	2,077	100
BY ECONOMIC CATEGORY	C	C'
- Developed countries with capitalist régime	1,500	72·3
- Third world	271	13
- Communist countries	306	14·7
World total	2,077	100

REFINING CAPACITY ON 1 JANUARY 1969		
Country	million metric tons per annum	%
1 United States	576	26·2
2 USSR	280	12·7
3 Italy	144	6·5
4 Japan	131	5·9
5 West Germany	112	5·1
6 Britain	97	4·4
7 France	93	4·2
Total for the seven countries	1,433	65·0
World total	2,202	100

Fig. 5 Oil consumption and refining capacity, 1969, by country and by region

In 1952 the United States Federal Trade Commission published a report which caused a stir, as the gist of its observations was that the cartel-like organisation of international oil affairs did not allow responsibilities to consumers to be discharged. Reports of European organisations also pointed out the damage done by the discriminatory price systems imposed by the large companies, showing that these systems were contrary to rational economic principles and harmful to European interests.[16]

Because of their enormous demand for oil, Japan and Western Europe must carry great weight in the hydrocarbons conflict. Their best manoeuvre in future is to take concerted action as buyers and play the old commercial game of potential purchasers who have money to spend but do not intend to use it foolishly.

Thus, one finds that there is complex interplay between the three main participants in this worldwide contest, namely the large international companies, the producer countries and the consumer countries. Spectacular advances, challenges between the participants and severe defeats have occurred. Although there were considerable alterations in the strategy of each of the participants during the 1960s, change was slow, with gradual reform usually prevailing over revolutionary ideas. But a turning point was reached in 1969–70. The buyers' market became a sellers' market. The negotiations in Teheran, Algiers and Tripoli during 1971 provided the first demonstration of the reversal of the previous state of affairs. World inflation, heavy demand for oil in Western Europe, United States loss of self-sufficiency in oil, the rapid growth of Japanese oil consumption, the increased effectiveness of anti-pollution policies and the political anxiety caused by the discord between Israel and its Arab enemies all contributed to the victory of the producer countries. Those countries took advantage of a favourable opportunity at the end of 1973 to combine withholding of supplies with stronger fiscal pressure. A new situation in oil affairs emerged. The essential difference may well prove to be attributable to the increased influence of a new participant in events – the state company.

Formation and development of state companies

Some governments have sought a means of reacting against the inflexibility of the international market and the persistance of a situation seriously at odds with the national interests of their countries. They have turned to direct participation in oil operations in the hope that this will serve for the purpose. Until now mention of oil operations has almost automatically conveyed the idea of the activities of the big international companies. But in recent times newcomers in the form of state companies have been entering the oil industry and become factors in economic and political considerations. These enterprises have been coming into oil affairs from both the supply and the purchasing ends, seeking to penetrate the system of triangular relationships between the producing countries, the consumer countries and the large international groups formerly serving as the unavoidable intermediary between the first two. The state companies resemble private concerns in that they, too, aim at the lowest possible production costs. But they have a public aspect as special instruments of oil policies of the exporting or importing regions. This book will devote particular attention to their formation, their development and their strategy.

But clearly the part played by a public enterprise must be viewed differently according to whether it belongs to a producer country or to a country with large consumption and little production. Although the Italian organisation ENI and the Algerian concern Sonatrach are given the same label, there are great differences between them. The same applies to the Spanish partially state-owned enterprise Hispanoil, the Libyan company Noc and the French ERAP. The term state company does not have the usual meaning in the context of oil.

Producer countries and state companies

Even within the producing countries one finds an astonishing variety of companies which all supply 'state oil'. This is already clearly apparent when one considers the Middle East, where Kuwait's state enterprise, KNPC, is very different from Petromin of Saudi Arabia. But there is an even greater diversity of situations in Latin America, where the creation of a public enterprise in the oil industry is customary.

In the Latin American environment oil consumption is still low and output is at a moderate level except in the case of Venezuela. Yet there are eight state companies, whose scale of operations ranges from near-monopoly to mere peripheral presence by comparison with the total oil interests of the country. The expansion of these Latin American state companies is often hampered for political reasons, which may vary considerably under successive régimes. All of these companies, together with the Mexican state company Pemex, are members of Arpel, the association for mutual assistance between the state companies of Latin America set up in 1954; but it has proved to be difficult to work out a coherent common policy within that confederacy. On studying the scope and volume of their operations one has the impression of looking into a kaleidoscope. Unlike Pemex, none of the eight companies has a complete monopoly of production, refining and distribution. Some of them, including Petrobras of Brazil, have the monopoly of production or, as Ancap of Uruguay does, of refining, or, as in the case of Enap of Chile, of both. But none of them has the monopoly of distribution. Few of them export crude or petroleum products but if they do, then the quantities are very small.

Just as the national policy concerning these companies varies greatly, so does the result of their operations, as table 1 shows.

Table 1

South America: state companies' production and refining capacity (thousand barrels daily)

	Year founded	Production		Refining capacity	
		1962	1967	1962	1967
Argentine:					
Yacimientos Petrolíferos Fiscales (YPF)	1922	268·6	312·3	217·0	263·0
Country total		269·0	314·5	365·9	433·3
Percentage of public sector		*99·9%*	*99·3%*	*59·3%*	*60·7%*
Bolivia:					
Yacimientos Petrolíferos Fiscales Bolivianos (YPFB)	1936	7·1	9·0	12·5	14·0
Country total		7·1	40·0	12·5	14·0
Percentage of public sector		*100%*	*22·5%*	*100%*	*100%*
Brazil:					
Petroleo brasileiro (Petrobras)	1953	91·5	146·0	242·0	309·5
Country total		91·5	146·0	298·6	366·1
Percentage of public sector		*100%*	*100%*	*81%*	*84·5%*
Chile:					
Empresa nacional del petroleo (Enap)	1950	32·0	33·0	44·0	92·0
Country total		32·0	33·0	44·0	92·0
Percentage of public sector		*100%*	*100%*	*100%*	*100%*

Table 1 continued

	Year founded	Production		Refining capacity	
		1962	1967	1962	1967
Colombia:					
Empresa Colombiana de Petróleos (Ecopetrol)	1951	28·8	29·2	36·0	71·6
Country total		141·5	193·4	79·2	130·8
Percentage of public sector		*20·4%*	*15·1%*	*45·5%*	*54·7%*
Peru:					
Empresa petrolera fiscal (EPF)	1934	3·8	6·7	1·5	21·5
Country total		58·0	66·6	55·8	94·0
Percentage of public sector		*6·6%*	*10·1%*	*2·7%*	*22·9%*
Uruguay:					
Administración nacional de combustibles alcohol y Portland (Ancap)	1931	–	–	45·0	45·0
Country total		–	–	45·0	45·0
Percentage of public sector				*100%*	*100%*
Venezuela:					
Corporación Venezolana del Petróleo (CVP)	1960	1·0	8·1	2·3	16·3
Country total		3,199·8	3,542·1	1,004·5	1,208·7
Percentage of public sector		*0·03%*	*0·2%*	*0·2%*	*1·3%*
Total state companies		433·0	544·3	600·3	832·9
Total all companies		3,798·9	4,335·6	1,905·5	2,383·9
Overall percentage of public sector		*11·4%*	*12·6%*	*31·5%*	*34·9%*

One barrel per day = approximately 50 tons a year

Source: *Petroleum Press Service,* June 1968, p. 216

It may be noted from the table that the output by state companies showed some increase over the years but their contribution to the national production was not yet large. In fact, the state companies' production rose by nearly 26 per cent between 1962 and 1967 (compared with an increase of 14 per cent for all the companies). Despite this good achievement their share of the total went up only from 11·4 per cent to

12·6 per cent. However, they were better placed in refining, because during this period their refining capacity expanded by about 36 per cent, against 24 per cent for all companies, whilst their share of the total rose from 31·5 per cent to 34·9 per cent.[17] But these figures are misleading, since they relate to Latin America overall, although in oil matters the totals are heavily influenced by Venezuela. The other countries are small producers with a relatively high proportion in the hands of the public sector. Venezuela, by contrast, is a very large producer (187 million metric tons in 1969, amounting to 8·8 per cent of world production) with a very small public sector. This may be seen more clearly by looking at Venezuela separately. The comparison between Venezuela and all the other countries is given in table 2.

Table 2

Public sector in Venezuela and in Latin America

	Production		Refining capacity	
	1962	1967	1962	1967
Total excluding Venezuela:				
State companies	432·0	536·2	598·0	816·6
All companies	599·1	793·5	901·0	1,175·2
Percentage of public sector	72·1%	67·6%	66·4%	69·5%
Venezuela:				
Percentage of public sector	*0·0%*	*0·2%*	*0·2%*	*1·3%*

However, changes are taking place in Latin America too. The state companies are being called upon for sizeable expansion. Encouraged by the victories of the Arab producing countries, oil nationalism is spreading rapidly and achieving successes. Since 1968 two countries with fairly promising oil resources, Peru and Bolivia, have nationalised the North American companies. In 1971 the government of Ecuador set up the state company Cepe and gave it considerable privileges in production, refining and transportation of crude oil. The recently installed President of Venezuela, Carlos Andrés Pérez, could anticipate oil revenue in 1974 totalling approximately four times the government's budget expenditure. He proposed setting up a special fund for domestic investment and 'advanced reversion' of the properties of foreign private oil companies because, he said, 'We cannot continue until 1983 without defining our oil policy'.[18]

In the Western Mediterranean, the Algerian company Sonatrach has managed by successive moves along the chain from sale to exploration to gain preponderance at the expense of foreign interests. Libya's enterprise, Noc, has followed suit later and is gradually expropriating the assets of foreign companies.

Consumer countries and state companies

Among countries which are small producers and large consumers there are few whose governments have chosen to equip themselves with state companies. Objectives in such countries are often of another kind, namely to adopt a body of legislation that safeguards the essentials of national interests, to encourage the build up of powerful private groups mainly funded with capital from investors who are nationals of the country and to regain abroad the unavoidable costs for the home market.

The European countries with Mediterranean coastlines have, however, taken a different line. Here the State has judged legislation and regulatory measures to be insufficient to assure the nation of a degree of independence in oil matters. The decision to become an entrepreneur in oil affairs has eventually been taken by each in turn. But the resultant state companies have produced dissimilar achievements.

The Spanish enterprise Campsa, for example, is a semi-state company but since 1927 it has been granted the monopoly of oil importing. The Italian ENI managed to develop remarkable refining capacity, adopting a bold commercial policy from the start, but has run into difficulties through not having succeeded in its efforts to find sufficiently worthwhile quantities of oil in exploration abroad.

France has undeniably managed to raise itself to modest success at international level, precisely because it has several strings to its bow. CFP, a partially state-owned organisation formed in 1924 under the Poincaré government with a 35 per cent state holding, has succeeded in entering the 'Club' of the big companies of the oil industry, achieving respect for its rights from British and American operators. Because of the determined competitive attitude of the enterprise, France did not suffer too badly from its shortage of oil before the Saharan deposits were found and became profitable. The company thus served to give the French oil interests an international base. However, the company was operating under arrangements which meant that it had to fall in with the wishes of the majors. In view of increased French needs for oil, the necessity of co-operation with an Algeria independent of French trusteeship and the aim of further prospecting in the franc zone, the French government decided to raise its stake in oil affairs. Accordingly, ERAP was formed at the end of 1965, by the merger of various state-owned concerns set up since 1939.[19]

Clearly, the oil policies of the three countries mentioned were not limited to launching and supporting enterprises in the public sector. The governments have always sought to combine respect of the interests in their national markets held by foreign companies with safeguarding the expansion of the state-controlled oil groups. In this there has been an interesting attempt to maintain pluralism, in reconciling the activity of international private capitalism with the promotion of state capitalism. But one may well ask whether the consumers of these countries have not had to pay the costs of this exercise.

State commercial oil holdings of Western Mediterranean countries

It so happens that the geographical unit of the countries either side of the Western Mediterranean is an oil world microcosm of especial interest to the observer, because tension between the various participants in oil affairs is noticeably severe there. The state oil companies are more highly developed there than elsewhere. It therefore seemed to the author that this region deserved special study, as a telling example of the way in which economic issues and political considerations have formed around a raw material whose importance to industrial development has become abundantly clear. This geographical area would appear to have its own distinctive features in the context of world oil matters. I shall endeavour to show how events of the 1960s have given the area characteristics of its own and have caused it to assume its particular position in international oil affairs.

Certain explanations are necessary in order to justify selection of this area and to mark out the scope of the study presented in this book. These observations will make it easier to clarify the plan adopted in presenting the material in this book.

The title indicates certain topics which must necessarily receive attention throughout if the book is to pay due regard to its avowed subject-matter. Four main bases for discussion will be used, namely the place in the economy, the time base, the geographical setting and the place in law.

Place in the economy; oil and other energy sources

Oil alone is under consideration in this book. True, it would have been interesting to present a combined study of oil and natural gas and thus produce a book on hydrocarbons in liquid and gas form. But in Europe the natural gas sector is becoming increasingly important, taking on a rôle of its own, independently of oil. The gap between these two energy sources in their place in world affairs is widening, as experience in the United States and in the USSR shows. Gas now has special individual characteristics which require separate study. It will, of course, be necessary to mention other energy sources, but only as the background to the persistent advance of oil as the main source of energy in the industrialisation of the countries of the Western Mediterranean basin.

Time base: the turning point, 1956–60

In order to understand the new features of the present situation and the importance assumed by the oil affairs of the states with Western Mediterranean coastlines it should be realised that until about 1956 Europe had suffered a shortage of oil that seemed to have lingered on ever since the Second World War. In many countries there were coal industry commitments limiting the advance of petroleum products. There was scarcely any free market for crude oil. In a situation of little surplus production and relatively feeble competition, supply and demand in a very limited market were regulated by the big international companies, which between them handled almost the whole oil output and owned a large proportion of the trading network.

However, since about 1956 economic and political upheavals in the oil world have been redoubled.

Change in the economic sphere has arisen from the increase in commercial supplies. In 1956 there were the first discoveries of economically viable reservoirs in the Sahara zone, at Edjelé and Hassi-Messaoud. 1956 also brought the first Suez crisis, causing the beginning of revolution in the tonnage of tankers. The combination of these two events gave rise to a kind of competition which until then had scarcely existed. North Africa very quickly assumed an important place in production, the start in Algeria being followed soon afterwards by the development of sizeable output in Libya. The second Suez crisis, in 1967, yielding the closure of the Canal, had the result of strengthening the strategic position of these two new producers in relation to the enormous consumer markets of Western Europe. The effects of the stiffer competition began to be felt. The majors were obliged to adopt more flexible price policies, together with more effective sales operations on the free market. At that time, too, post-Stalin Russia returned to the market outside the Communist bloc, considerably increasing its export capacity in order to obtain the earnings needed to finance Soviet development and improvement of the standard of living of the republic's population. It should be added, moreover, that the restrictive policy introduced by the federal authorities of the USA virtually excluded the greater part of Middle East and African oil output from the expanding US market which represented about 40 per cent of the world market. New trade channels were set up to deal with the increased demand in Western Europe; overall factors of equilibrium which had long been paralysed were restored, especially as large tonnages of 'orphan' crude oil provoked greater competivity in the market.

Events in the political sphere brought profound change in overall features of oil strategy in the countries of the Mediterranean basin. Algeria gained independence and sought to free itself from guidance by its former trustee, whilst discussing with France the special relationships between the two countries. In Libya the discovery of oil in vast stretches of the desert immediately gave that country a place in international oil affairs. Moreover, the various oil-producing states adopted the notion of joint action on behalf of their interests and set up OPEC in 1960. First Libya then Algeria decided to join that organisation.

In the consumer countries further moves to set up or to strengthen state companies were proceeding at the time when the Treaty of Rome, in 1957, indicated new prospects, calling for the opening up of frontiers, increased trade and greater competition. The Treaty gave fresh impetus to oil dealings between the two sides of the Mediterranean but required this to be viewed in the larger context of Europe and Africa and of the distribution of trade throughout the world. The entry of Britain into the Common Market placed still greater emphasis on the broadening of international trade. The 1971 and 1973 crises have broken the special relationships between the two sides of the Western Mediterranean but they have served to point out more clearly the special features of the 1960s for consumers and producers there.

Thus, oil politics in the Western Mediterranean countries take place in a wider context which must be borne in mind, otherwise the new features of the situations which will be discussed in this book may not be understood.

Geographical setting: the Western Mediterranean

The new importance of the area derives from the facts already mentioned. The consumer countries which will be studied are Spain, Italy and France, which are the only countries of Western Europe with state oil enterprises in the full sense of the term, that is, wholly state-owned.

As regards the producing countries, the geographical area under consideration involves Algeria and Libya, also Morocco and Tunisia as minor producers. All four of these countries have state-owned or state-controlled oil companies, although the Moroccan and Tunisian companies are of little importance because of the low level of oil output. By contrast, Sonatrach of Algeria is growing very rapidly and the Libyan enterprise, which was originally Lipetco, then Linoco and is now Noc, is expanding every year, following in the path of Algeria's experience during the 1960s.

Between producers and consumers and between the state companies of each of these countries a set of arrangements has grown up. Care must be taken to observe correctly the spider's web of joint interests recognised either side of the Mediterranean. But since some of the countries are producers and others are consumers there is here, as elsewhere, an interplay of supply and demand and of measures to reconcile interests which are divergent at the outset. Here, more than elsewhere, politics seeks to achieve a *modus vivendi* among opposing interests.

It is a function of politics to achieve the harmony indicated by geographical proximity and politics can play some part in bringing local loyalties into operation, but politics cannot overcome certain limitations imposed by the basic economic realities of the case.

Place in law: the criteria defining a state company

One of the levers used by nations for applying their political aims in oil matters is the formation or the strengthening of state companies. Both the producing and the consumer countries of the area under consideration adopt this means. On the one side of the Mediterranean there are Sitep of Tunisia, the Moroccan BRPM, Algeria's Sonatrach and Noc of Libya; on the other side Hispanoil, ERAP of France and the Italian ENI are comparable to each other. The setting up of state oil enterprises is not peculiar to the Western Mediterranean; it occurs also in Latin America, the Near East and the Middle East but it is more prominent and more effective in the countries of the Western Mediterranean. The novel features of this development have usually received little attention from observers. I propose to fill this gap.

But what exactly does a state company mean in the context of oil affairs? Does the term apply only to a company with public capital, namely a wholly state-owned company? If so, what place does one assign to partially state-owned companies, for example BP and Compagnie Française des Pétroles, which are more indirectly state-controlled? They cannot be disregarded, although consideration of them is relatively marginal for the purposes of this study.

The state oil sector is a remarkable phenomenon because it is truly paradoxical: usually it does not have a genuine monopoly, therefore it is in a competitive situation. A state oil group lacks an initial investment of private capital to enable it to obtain a position in the market dominated by large companies with ample funds; therefore public funds have to be assigned to it. Even so, as Pierre Guillaumat, Chairman and Managing Director of ELF-Aquitaine has pointed out in the language of French legal terminology, it is an industrial and commercial concern which has to make its way in a competitive situation. Yet undeniably the raison d'être of a state concern is to act as an instrument of economic policies thought to be in the interests of the country to which the concern belongs. A company of this kind is placed at a junction of politics and economics or a crossroads of economic rationality and political design: it is situated at a position where day to day needs come together but are not necessarily in agreement and may even be irreconcilable. If a state company serves mainly to test out economic theories of profitability, why not entrust this to concerns in the private sector? If a state company is too political is it not economically too costly to be really in the national interest?

It will be necessary to give a personal assessment of the state companies' successes and shortcomings and of the effectiveness of oil policies adopted in the countries of the Western Mediterranean in this period of market upheaval marked by the dynamic progress of state participation in oil affairs. When the results are weighed in the balance sympathy must give way to strict judgment of the facts. All the same, it is necessary to preface this critical analysis by an attempt to understand the aims pursued by the consumer and the producer countries on each side of the Mediterranean according to their particular interests.

Notes

[1] E. Mattei, *Problems of energy and hydrocarbons,* ENI, Rome 1961, pp. 13 et seq. From a speech given on 8 January 1959 in Rome.

[2] M. Byé, 'La grande unité interterritoriale', *Encyclopédie française,* vol. IX, *L'univers économique et social,* Larousse, Paris 1960, section B, chapter 3. The turnover of Esso International for 1972 was 22,070 million dollars, from crude output amounting to 12 per cent of world production.

[3] Ibid. In 1968 the share of the majors in United States production had risen to 38 per cent. This slight increase in percentage does not alter the essential facts shown by the 1949 data.

[4] J. Masseron, *L'économie des hydrocarbures,* Ed. Technip, Paris 1969, pp. 16–17. The eight leading companies are Standard Oil (New Jersey), Royal Dutch Shell, Mobil Oil, Texaco, Gulf Oil, Standard Oil (California), British Petroleum, Compagnie Française des Pétroles.

[5] M. Byé, op.cit.

[6] *The Petroleum Economist,* January 1974, p. 20.

[7] 'The American oil dilemma', *The Economist,* 28 February 1970, pp. 62–3.

[8] P.H. Frankel and W.L. Newton, 'The state of the oil industry', *National Institute Economic Review,* September 1960, 11, pp. 16–25.

[9] All these details are from E. Mattei, op.cit., pp. 15–16.

[10] H. Melan, 'Le pétrole, mythe politique?', *Projet,* March 1970, 43, p. 254.

[11] On all these points see J. Chardonnet, *Géographie industrielle,* vol. 1, *Les sources d'énergie,* Sirey, Paris 1965, pp. 326–8.

[12] M. Laudrain, *Le prix du pétrole brut, structures d'un marché,* Médicis, Paris 1958. See also J. Masseron, op.cit., p. 26.

[13] Sr Betancourt has written a number of books and articles on the combination of economic considerations and political considerations in oil matters.

[14] *Problèmes économiques,* 29 September 1964, 874, p. 13.

[15] *Petroleum Press Service,* March 1969, p. 92. Common Market imports in 1968 were 315·5 million metric tons, whilst internal production was just over 14 million metric tons. In 1968 Libya provided 42 per cent of West German oil imports, 23 per cent for Italy and 14 per cent for France. France obtained 25 per cent of its oil imports that year from Algeria.

[16] M. Byé; *Journal Officiel* (France); Recommendations and Reports of the UN Economic and Social Council, no. 15, 1954; 'The International Petroleum Cartel', Report of the Federal Trade Commission to the United States Senate, 1952; 'The price of petroleum products in Western Europe', Report of the UN Economic Commission for Europe, 1955; P. Clair, 'Le rôle des Etats, du secteur public et des entreprises privées dans l'économie pétrolière', *Les Annales de l'économie collective,* 1, January–March 1966.

[17] Statistics from *Petroleum Press Service,* June 1968, p. 216 et seq.

[18] *The Petroleum Economist,* April 1974, p. 148.

[19] *Revue française de l'énergie,* January 1966, p. 237. Erap inherited the rights and obligations of Rap and BRP, which had carried the burden of the national oil effort from 1945.

1 Aims of the Consumer Countries

Ever since oil began to be an important commodity all nations have sought assured supplies safe from political risks and at the same time supplies at reasonable cost in terms of both price and currency. Three countries, the USSR, the United States and Britain, have succeeded in finding a solution to the oil problem.

Output within the USSR in 1973 was 421 million metric tons, amounting to 15 per cent of the world total. This level of output is higher than internal consumption, leaving a surplus for export. Undoubtedly there are untapped oil reserves within the vast areas of the Soviet Union but their exploitation would require heavy investment, after precise calculation of comparative returns. Accordingly, the future of Soviet oil supplies depends on the rate at which the USSR sets out to develop oilfields in its own territory and on the distance of those oilfields from the present industrial centres.[1] For the foreseeable future the Soviet Union will no doubt be obliged to pursue a policy of importing oil. It has already begun to import oil from the Arab states and the motivation for importing would be to secure development within the Soviet Union whilst maintaining a firm hold on established Soviet export channels in the Communist bloc and whilst making provision for foreign currency receipts from oil exports on the free market (some 50 million metric tons a year).

The United States, the world's leading oil producer, had a crude oil output in 1973 of 513 million metric tons, which was 18 per cent of world production. But US demand in the same year was about 806 million metric tons i.e. more than 29 per cent of world consumption. The gap of about 25 per cent in 1971 widened to 30 per cent in 1973 and was expected to reach 35 per cent in 1974. Before the 1973 embargo the share of the difference attributable to Arab oil was 10 per cent. Thus, with consumption increasing whilst output was tending to remain steady or even decline, the US oil balance was seriously in jeopardy by 1974. Despite the new oilfields in Alaska, reserves at the beginning of 1974 were only 5·5 per cent of world reserves and less than 10 years' production. However, the United States was undoubtedly preparing to adopt a more vigorous approach to the oil problem. Oil production from coal and from

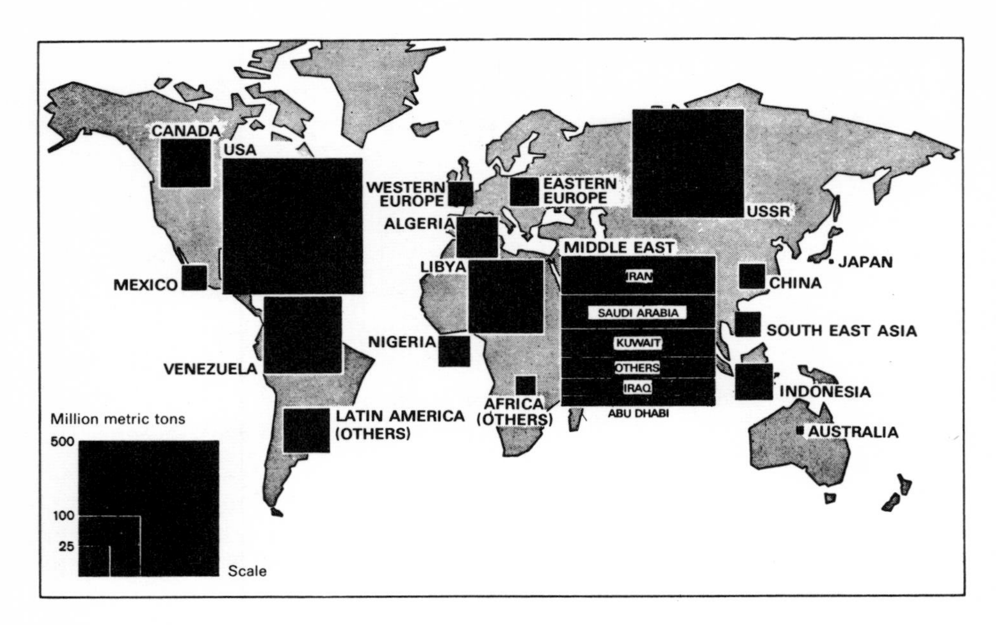

Fig. 6 Crude oil production, 1969

Source: Comité professionnel du pétrole

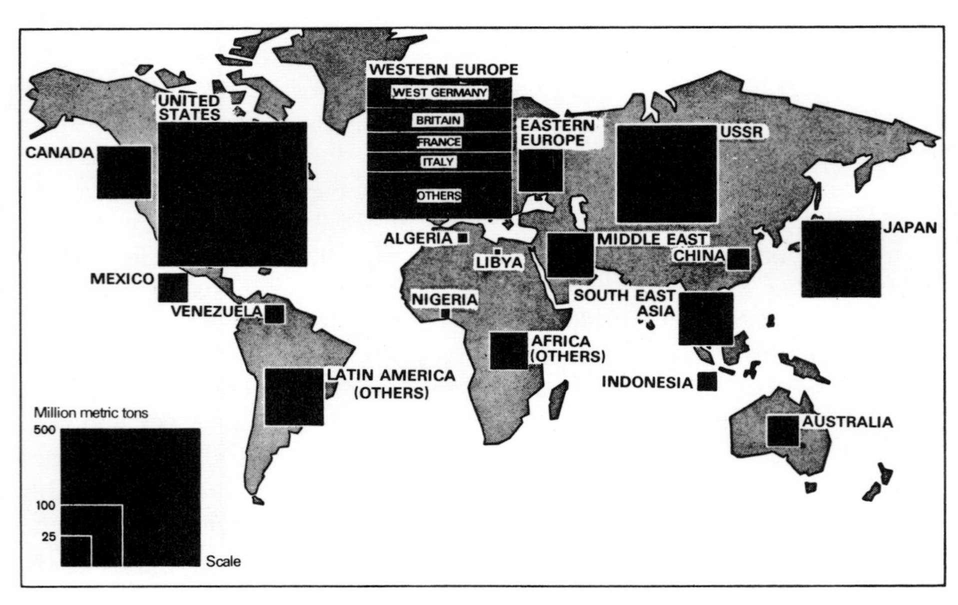

Fig. 7 Crude oil consumption, 1969

Source: Comité professionnel du pétrole

shale was rendered economically viable at the level of oil prices reached by their trebling in the second half of 1973. If a major programme of investment in domestic oil production is undertaken – and there is every indication in 1974 that this has been set in train – the United States could regain self-sufficiency in oil early in the 1980s. Precise economic calculations showed that the costs of determined national policy in this connection, although heavy, would be lower than expenditure on imports from abroad (at a total of upwards of 26 thousand million dollars from 1980 onwards).

Britain, too, was working on an eventual solution to the oil problem, although along completely different lines. In 1974 it was still importing the whole of its domestic requirements (122 million metric tons, equalling 4·5 per cent of world demand) but from the outset it managed to acquire extremely rich oilfields in various parts of the world, especially in the area which is most productive and which has the largest reserves for the future – the Middle East. British oil holdings have been described as the last remnants of the illustrious British Empire. Britain supplies oil to many countries and is an astute trader in this commodity. Through its world policy and as a result of the power of Shell and BP it has more than made up abroad the costs it had to accept at home. Now the recent discoveries in the North Sea enable Britain to take an optimistic view of the future. Britain will be self-sufficient in oil in 1980, because according to 1974 estimates the British share of offshore fields in the North Sea should be yielding between 100 and 150 million metric tons by the beginning of the next decade. Dr P.R. Odell is one expert who believes that Europe could reach self-sufficiency in oil as early as 1985, even without exhaustive exploration of the surrounding seas. However, the 1973 crisis and the resultant rise in crude oil prices gave a tremendous impetus to exploration and development of oilfields in European territorial waters. Nevertheless, Britain faces special difficulties in the second half of the 1970s because the exploitation of oilfields was delayed by the economic depression and the social crisis prevailing in the United Kingdom in 1974. Oil production was expected to be fully restored in 1976 but 1975 yields were estimated at no more than about 5 million metric tons, instead of the 25 million metric tons that had been hopefully anticipated.

Having looked at these three differing examples of success, let us now consider the positions of Spain, France and Italy, which have scarcely any natural oil resources in their national territory. Given this initial handicap, what policies may this powerful consumer group be expected to pursue?

There are considerable disparities between France, Spain and Italy in their economic circumstances. They have dissimilar levels of development and widely differing rates of industrialisation. Nevertheless these three countries have some common features in the matter of liquid fuel, including the absence or near-absence of oil in their territory, the necessity of importing large quantities in order to support domestic development, the need for security of supplies in order to protect the functioning of the economy from risks arising abroad, the endeavour to minimise the effect of these imports on the balance of payments, and the desire to free oil affairs as much as possible from the influence of foreign interests and from decisions taken in centres of power outside their national territory.

As a result, the existence of some objectives common to all three countries may be noted. The degree of resoluteness adopted in pursuing those objectives varies from one country to another but none of these countries neglects them entirely. Such differences as may be apparent arise mainly from the way in which the national policy of each country combines specific considerations. Moreover, West Germany and Japan suffer from similar imbalances. If the author is naturally inclined to pay most attention to the case of France, the choice of France for reference is justified by the fact that France has pursued for a longer period of time and more consistently the aims common to all three of the consumer countries under special consideration in this book.

Those aims are diverse and they evolve in the course of time. For that reason the author will do scant justice to a theme which is tending to become an important article of national policy – that of *combating pollution and safeguarding the environment.* Although the employment of liquid fuel rather than coal has been favoured precisely for reasons of greater cleanliness, the introduction of new technology is invariably accompanied by a great deal of misuse. Thus, recent legislation has sought to lessen the noise of internal combustion engines, to reduce the lead content of petrol, to prevent oil tankers cleaning their tanks at sea, to stop pollution of beaches and to obtain better siting of installations. Already 10 per cent of total capital investment in refineries is spent on combating atmospheric and water pollution. But no real progress will be made until there is acceptance of the need for an international 'authority for the environment' carrying equal weight with other authorities at supranational level and capable of obtaining obedience from transnational companies. The endeavour to protect the environment will not be given its rightful place in the consumer countries until two prerequisites have been achieved – the end of the present situation of energy shortage, or rather of panic about energy resources, and a shift in collective thinking towards treating economic cost in itself as being one item in the sum of overall cost to society. Hence, that factor has a place in the general context of policies common to the consumer countries, among the extremely diverse set of considerations involved. Despite the diversity it seems possible to distinguish two overriding issues – the aim of security of supplies and the aim of minimum cost.

Aim of security of supplies

The aim of security of supplies is the most prominent of all in national oil policies. The necessity on the part of the consumer countries to obtain a guaranteed minimum level of oil supplies has become all the more urgent as a result of the decline of the coal-based economy and the increasingly large part played by oil in energy demands. In 1900, when total world oil production was 20 million metric tons, this was a minor consideration. But in 1950, when the overall level had risen to 500 million metric tons, this had become really important. It must inevitably become increasingly vital as levels of consumption continue to soar. In 1960 total production passed the thousand million metric ton mark. By 1974 two million metric tons has been reached and there is little sign of any slackening. Imported oil now supplies more than 50 per cent of Europe's energy needs and the figure is expected to rise to between 60 and 65 per cent in the mid-1970s.

The European countries are in danger of becoming dependent on a small number of oil-producing countries. A few statistics suffice to demonstrate this fact. In the period from 1950 to 1968 there was a sevenfold increase in world crude oil production. Over the same period crude oil imports by the six EEC countries rose from a total of 27 million metric tons to 317 million metric tons in 1968, which is an almost twelvefold increase. Hence, the energy problem is increasingly bound up with that of oil supply.[2]

France, Italy and Spain have no domestic oil production and therefore need to be able to rely on foreign sources for continuity of supply to sustain their economic growth. In order to sell national labour input in finished products these three countries must be assured of regular fuel supplies. In the complex conditions of a highly automated economy this is a vital consideration. The matter becomes urgent in time of war or crisis but it is already of great importance in peacetime. Therefore the constant aim is to come as close as possible to complete security of supply.

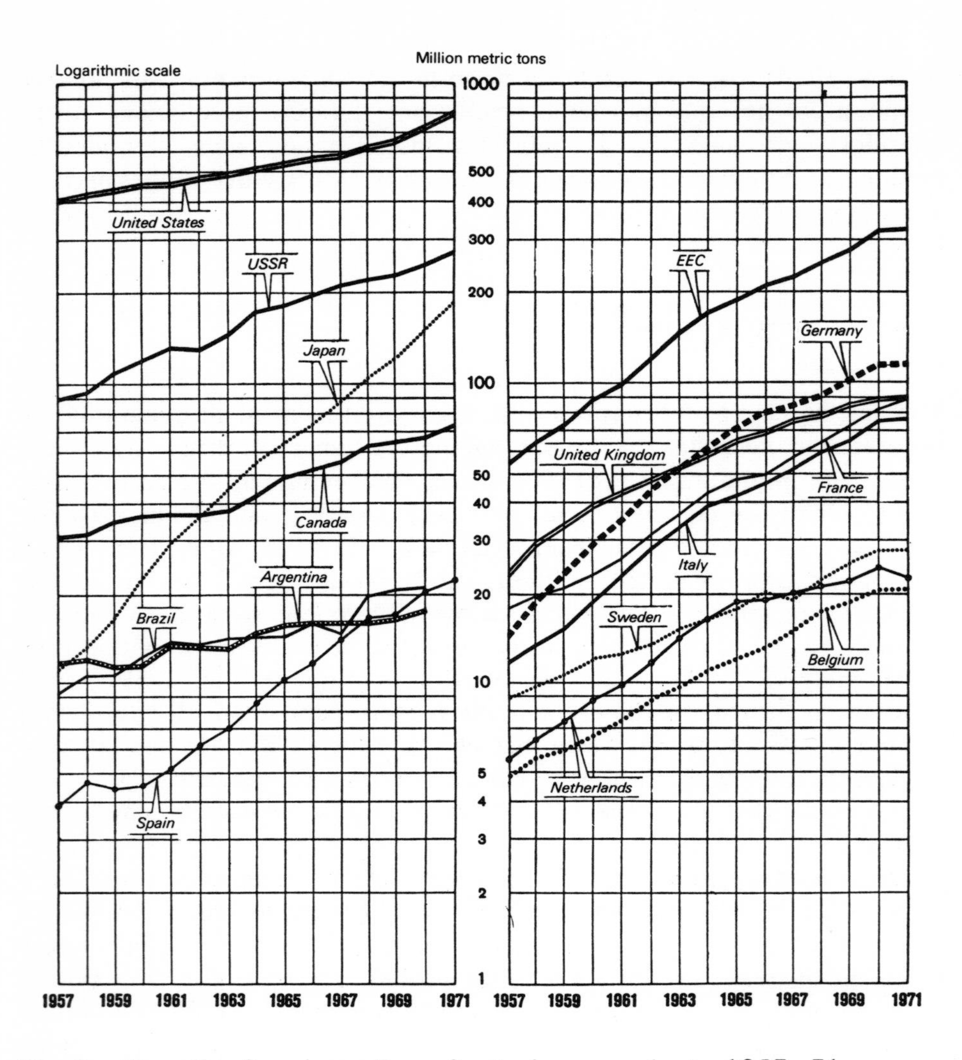

Fig. 8 Growth of consumption of petroleum products, 1957–71

Source: Comité professionnel du pétrole

Security of supplies in wartime

War compels realistic thinking on the supply problem. In wartime survival depends on obtaining sufficient supplies; the decisive factor becomes the ability to overcome the risk of destruction. No country can have an efficacious and credible national defence policy if the receipt of supplies is not assured – and this is all the more imperative when its land, air and sea forces are involved in continuous movement. Victory by a nation is linked with its fighting power, but weapons only serve their purpose if systematic refuelling is carried out. Hence, there is a whole area of oil policy concerned with military considerations.[3]

At the beginning of this century one man more than any other perceived the strategic importance of oil. He was the First Lord of the Admiralty at that time, Winston Churchill. In 1913, when he was 38 years old, the British Cabinet entrusted him with the task of overcoming the threat posed by the huge naval rearmament programme introduced by the Kaiser in the years immediately preceding the outbreak of the 1914–18 War.[4]

Jean Baumier writes:

> At that time the British Admiralty was faced with two major problems – the insufficient gun power and the inadequate speed of its vessels. The solution to the first problem was provided by equipping the Royal Navy's vessels with 15-inch guns. The second problem was overcome, after a great deal of controversy, by a technological transformation in the motive power of the vessels: coal was replaced by oil and the obsolescent boilers were replaced by huge engines.
>
> Winston Churchill has explained, in the volumes of 'The World Crisis', how this changeover by the Navy to oil-fired engines gave a 40 per cent greater radius of action, enabled a fleet to re-fuel at sea and made it possible to realise the high speeds in certain types which were vital to their tactical purpose.
>
> In order to equip the cruisers, battleships and submarines of His Majesty's Fleet with oil-burning engines sufficient quantities of oil were needed. This specific problem of supply was studied by Lord Fisher and at the express instigation of Winston Churchill arrangements with the Anglo-Persian Oil Company proceeded under the Chancellor of the Exchequer, Lloyd George. The negotiations culminated in the signature a year later of an agreement giving the British Government a majority holding in Anglo-Persian. The Government provided the company with the funds required for the exploitation of its oilfields in Persia. In exchange, the British Navy obtained an assurance that henceforth it would receive all the fuel that it needed, whether in peacetime or in wartime.[5]

Churchill was able to obtain the passage by Parliament of the necessary Bill just in time for the agreement to be ratified a few days before the outbreak of war. For £2 million the Admiralty became the majority shareholder in Anglo-Persian, later re-named Anglo-Iranian. The present British Petroleum is the successor to that original company and the Government still has a controlling share in the capital, although now reduced to slightly less than 50 per cent.

This example shows how a technological revolution, namely the change over from coal to fuel oil in time of impending war on a large scale, obliged a government to seek at all costs to obtain security of its supplies. Faced by the threat from Germany, the British Cabinet acted quickly to take over control of a private company in order to make more certain of wartime supplies to the Navy. It often happens that new organisations designed for wartime purposes subsist in altered form in peacetime. This change of the use of a private enterprise under pressure of strategic and political necessities offers a lesson which should not be overlooked by posterity.

But this event, which has attracted public attention, should not be allowed to obscure another. Whilst Britain's security aims were achieved through the government's decision to take over a private company, the success of the venture was due to the part played by the Royal Navy which, fortified by its imperial traditions, managed to give relative protection to the tanker fleet throughout the war.

France, for its part, had no such bold policy on the eve of the 1914–18 War. And no leader arose, as Winston Churchill did, to point out the strategic importance of oil in peacetime and the even greater need in wartime. The amounts imported into France prior to 1914 were so small that the idea of assuring supplies did not occur to anyone in a position of influence. The problem was discovered during the hostilities by Clemenceau himself. Requirements for a war which made ever-increasing demands on industrial resources obliged the policy-makers to try to deal with all the oil problems at once. In order to obtain security in refuelling it was necessary to rationalise transport, storage and distribution arrangements; it became indispensable to make careful investigations of the cost of effective national defence; it seemed imperative to find means of making measures in the national interest prevail where there was insoluble conflict with private interests.

During the first three years of the war the State refrained from intervening too forcefully. It left the task of supplying the French market to private concerns, whilst exercising some control to ensure that military needs were given priority for imports and stocks. But with the growing mechanisation of the war and the constantly increasing precariousness of shipping because of the deadly activity of German submarines in the North Atlantic, there was no alternative to the adoption of a more ruthless policy. This new line in French policy, which had at last grasped the vital importance of refuelling, was expressed in terms worthy of a war communiqué in a personal letter which Clemenceau, as Head of State, sent to President Wilson on 15 December 1917:

> At the decisive moment in this war, when the French Army is about to commence vital military operations on the French front, the French forces should not at any time be in danger of running out of the fuel required for the trucks, the aircraft and the motorised field artillery. Any breakdown of petrol supplies would cause the sudden paralysis of our forces and could compel us to accept peace on terms unacceptable to the Allies. . . . Now, there is a risk of vital stocks being completely exhausted if measures are not instituted and carried through to completion by the United States. For the sake of the Allied cause such measures can and must be taken without a day's delay, if only President Wilson will obtain the supply by the American petroleum companies of the reserve stock of 100,000 metric tons being held in tankers, which is necessary for the armed forces and the population of France. If the Allies do not wish to lose the war, France embattled at the height of the German onslaught must possess the fuel which is as necessary as blood in the coming battles.[6]

Spain and Italy did not have to face problems of survival at this early date. Spain was not a combatant in the First World War and Italy was only marginally involved as a support force on the Allied side.

But in the Spanish Civil War the key factor contributing to final victory of the Falange over the Popular Front was that the Falange was constantly supplied with liquid fuel by powers sympathising with its cause.

The problems recurred in the same way in the War of 1939–45 but even more acutely because weapons and transport moved at greater speed. However, in that war France and Italy had limited engagement on the battlefields and Spain did not take part. The annexation of Roumania, the Wehrmacht drive eastwards and the attempt to transfer the scene of conflict to the Near East contributed to the difficulties of the Axis forces in obtaining supplies for refuelling, despite the massive stockpiling before 1939 and the successes in making petrol substitute. Italy was only able to hold out through German assistance; after the defeat in 1940 Gaullist France was only able to take a small part in the final victory through the assistance in liquid fuel supplies given by the Allies.

The lessons of those wars will not be forgotten. In order to win a conflict engaging such massive consumption of liquid fuel it has become indispensable to be certain of avoiding any shortage of that vital commodity.

Nowadays the losing side will be the one which has not succeeded in making effective arrangements on a worldwide scale to ensure that it will receive energy supplies. Atomic capabilities will make little difference in this matter, at least in the early stages of war. Nuclear weapons will be used only as a last resort at the end of a process of escalation which will require the prior use of conventional forces. In principle they exist in order to deter the aggressor from using them first. But their use might possibly be justified if either side attempted to seize control of, or merely to paralyse to a large extent, the oil resources required for the survival of the opposing forces. Clemenceau's dictum still holds good: wars consume greater quantities of liquid fuel than of blood.

Security of supplies in peacetime

The first essentials in the wartime economy, especially that of liquid fuel supplies, have been discussed at some length above. Comparison of the British policies, thoroughly worked out before the trial of strength in 1914, with the French expedients largely improvised in the midst of the drama, is instructive in several ways. But what is true in wartime is also true in peacetime if there is a valid principle that, to paraphrase Clausewitz, peace is the continuation of war by other means.[7]

In peacetime a nation's mastery of its fate, which is one aspect of security, implies independence, or at least comparative power of economic and political decision. Accordingly, every State applies the lessons it has learnt in time of conflict and endeavours when peace is restored to achieve oil companies directly or indirectly under its control. It seeks above all to be less dependent on foreign interests for its supplies; it also strives for supremacy in its own territory, that is, to escape the clutches of the international companies by encouraging or creating within its territory enterprises that are able to make decisions independently of foreigners. These enterprises in the state orbit are indeed designed to be fundamental interests of the policy of security of oil supplies which every State plans to pursue.

But manifestly the security aim covers a larger area than the setting up of state-controlled companies. In fact, it spreads over into the domestic economic policy and the international politics of every sovereign State, because at the last analysis the State is responsible (or is believed to be responsible) for the regularity and continuity of supplies. The State is blamed in the event of breakdown of oil supplies. It is the State which is called to account. By its attitude in international affairs the State favours or obstructs the operations of private enterprise. The quest for security thus becomes directly expressed in political terms. It is reflected in the overall behaviour of every State and in its trading policy. In power relationships at international level this is a weighty consideration. Attempts to diversify sources of supply in order to lessen the hazards which might be caused by events and to reduce the political risks; the endeavour to maintain cordial relations with the limited number of producing countries; the discreet support of policy by means of conventional diplomacy; government support to keep foreign markets for the commodity strong; the inclusion of reciprocal advantages in oil negotiations with suppliers abroad; efforts to avoid the development of too great a dependence on any one business partner: all these practices may be recognised as being reflections in everyday experience of the vital need for security.

Each in its own way and with differing degrees of intensity, France, Italy and Spain are indeed pursuing this same design. That is the explanation of Mattei's efforts in the Middle East and his trading flirtation with the USSR. To that end France prospected at numerous sites on the mainland and in French-speaking territories before adopting a determined policy of worldwide exploration from 1960 onwards. In order to obviate any risks of pressure, or even the mere possibility of extortion, from their Algerian partner, the French embarked on a policy of spreading their risks throughout the world and not depending on any one country of supply which might be tempted to set its terms too high. Similarly Spain, through the partially state-owned company Hispanoil, has laid its plans for safeguarding its interests in hydrocarbon supplies; yet some inflexibility in its foreign policy towards Arab states has been noted.

The effects of the need for security are not solely international. There are smaller effects in domestic affairs. Governments do not hesitate to regulate the market itself, in order to maintain a certain degree of price stability, if either military or civil considerations so require. Usually such moves oblige enterprises in the trade to keep stocks and build up reserves. The object is to prevent national defence facilities and industries from being held to ransom because of market shortages or from suffering as a result of sudden unavailability of essential supplies (oil gas, petrol, fuel oil, lubricating oil, etc.).

This policy of seeking security of supply which has been practised separately by each consumer country has been raised to the level of a concerted European aim.[8] The nine EEC countries seek to adopt a common stockpiling policy in order to safeguard supplies. There have been differences of opinion between the Member States on the safety margin considered necessary. In line with its normal planning, France has advocated stocks equivalent to three months' mean domestic consumption in the previous year. The Italians have favoured less stringent precautions and in 1974 their point of view still prevailed. An EEC Directive dated 20 December 1968 imposed an obligation[9] to hold stocks equivalent to 65 days' mean domestic consumption in the previous year, with the option of reducing this amount by 15 per cent in the case of a Member State engaged in exploiting oil reserves in its own territory.[10] The crisis of the winter of 1970–71 speeded up EEC consideration of the matter, as well as providing justification for the French assessment, so that study of more drastic measures was undertaken. Indeed, the Commission proposed to the Council that the 1968 Directive on stockpiling should be amended to raise obligatory stocks of the three most important product categories to a level corresponding to 90 days' consumption in the previous year. In addition, consideration is being given to the matter of requiring importers of hydrocarbons to supply confidential information to the EEC Commission on their import programmes for the following year. That measure would

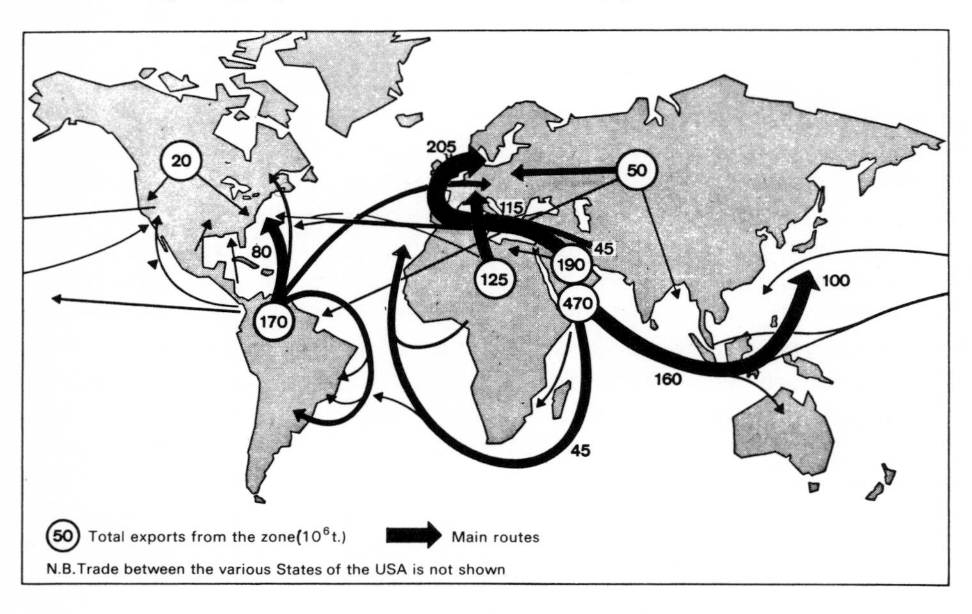

Fig. 9 Main supply routes in 1967 before the Suez crisis (crude oil and refined petroleum products)

Source: *Oil and Gas Journal*

enable effective action to be taken to prevent possible shortage and would permit efficient planning of oil purchasing. Here, again, there is an objective of obtaining security of supply but this time on the scale of a large area of Western Europe that has discovered the strength which a concerted purchasing policy gives.[11]

Security of supplies in the event of crisis

Although widespread war is unusual, peace throughout the world is equally rare. All too often there are crises which inevitably affect price levels and supplies. The most vulnerable region of the globe, precisely because of the oil riches in the subsoil, is the Middle East. Europe has experienced two serious emergencies connected with the political instability of the oil-producing countries east of Suez, the first in 1956, the second in 1967.

Admittedly, France and Britain were largely to blame for aggravating the conflict during the 1956 crisis. After Colonel Nasser had seized control of the Suez Canal and after the IPC pipelines had been destroyed those two countries found their supplies of Arab oil suddenly cut off. This vital liquid continued to flow from Aramco outlets but with a boycott imposed on the British and French. On the other hand, Germany, Italy, Spain and, of course, the United States were still able to obtain supplies and scarcely suffered from these retaliatory measures. But at that time France and Britain had to appeal to the Americans for assistance, to prevent a sudden complete stoppage of supplies. With the Canal out of use, oil had to be shipped by the much longer route round the Cape of Good Hope. At that time few supertankers were in existence to enable extra tonnage to offset the cost of the greater distance of the voyage. Therefore the oil supplies came from the continent of America, above all from Venezuela. The distance from the Gulf of Mexico to Europe was half the distance from the Persian Gulf to Europe via the Cape. American tankers were re-routed and the efficiency of the well-organised trading networks of the big international companies was of great value. The necessity of introducing partial rationing had to be accepted but there was occasion to discover the justification for stockpiling policies and the benefit of those policies. There was an acute feeling of insecurity but the united front of the producing countries finally crumbled. The normal flow of trade with the Middle East was very quickly restored and the building of bigger and bigger tankers was undertaken in order to make the detour round South Africa profitable. Indubitably the experience caused political leaders to swear that they would never again get into such a desperate

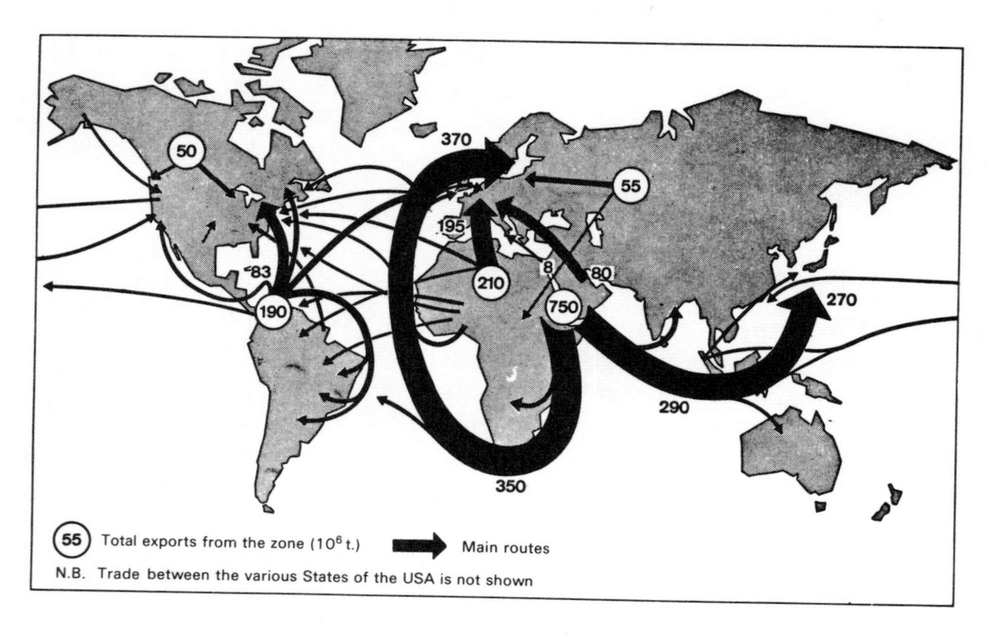

Fig. 10 Main supply routes in 1971 (crude oil and refined petroleum products)

Source: *Oil and Gas Journal*

situation. Relations between France and Israel became troubled and under the Fifth Republic the need for security of oil supplies became an important reason, although not the only one, for rapprochement with the Arab countries.

Therefore the second emergency did not assume dramatic proportions for France, and still less for Italy and Spain. The changed situation could be attributed to political reasons, especially the pro-Arab attitude of the French Government. At the same time Spanish and Italian policy remained essentially unchanged in a friendly attitude to the oil-producing states of the Eastern Mediterranean and the Persian Gulf. However, international trade channels had changed greatly since 1956. European dependence on the Suez Canal had been declining: 80 per cent of European supplies used that route in 1956 but only 40 per cent in 1967. Moreover, this time the embargo was limited: France under General de Gaulle was spared retaliatory measures. And, in addition, oil output in 1967 came from a greater number of producing countries. A new factor of tremendous importance was that sizeable production had grown up in some new areas which were all west of Suez – Algeria, Libya and even

Nigeria. Algeria and Libya had the advantage of being well placed in relation to the big European consumer markets. In 1967 they were already sending more than 75 per cent of their output to Europe.

Although the crisis did not prove to have any great effect on the three consumer countries under special consideration in this book, it again demonstrated to governments and pointed out to public opinion the importance of the aim of security of oil supplies. In wartime or in peacetime, or in periods of crisis more or less occupying the interval between those two circumstances, any country which is a major consumer without domestic oil production must, in order to live, or even survive, at all times avoid provoking a sudden stoppage of its oil imports. This is all the more essential with the growing proportion of use of oil as a source of energy. Because it has become economically vital, oil has acquired such a degree of political importance that no nation's diplomacy can treat it lightly.

Maximising security of supply

Commenting on the 1967 crisis, Professor Adelman pointed out in a report that, 'The extreme of the security problem is clear enough: be prepared for a total cessation for a limited period'.[12] If this could be achieved, then security was almost complete.

The American expert went on to consider how little the consumer countries can do about the risk of a sudden stoppage of oil supplies. A new petroleum area might be brought into production but this was too slow a process for the problem posed, that of cessation of supplies. It was also possible to diversify the risks, by obtaining supplies not from one country but from several; but the shutdown introduced in the event of a crisis was a concerted action. Moreover, oil collection was usually performed by the large international companies and there was extreme distrust of these in time of international tension. A state company could not provide security of supply any more than a privately owned company could. Indeed, it was an even more obvious and more tempting target because of its more political nature. But public or private ownership was simply irrelevant to the chances of a concerted shutdown. For Professor Adelman the essential point was in the inconsistencies of European energy policy and especially in its refusal to make a clear choice between coal, which he condemned, and oil, which was more economic than ever.

Calculating the cost of keeping the coal industry in operation, Professor Adelman stated that in 1966 coal production in the OECD countries of Europe was about 212 million metric tons oil-equivalent, excluding coking coal. He said that the total replacement of this tonnage by oil would have saved 3·5 thousand million dollars. He therefore recommended that instead of 'the dead weight loss to Europe' there should be 'an adequate security program by stockpiling crude oil'. Adding up all the costs, under the conditions of the mid-1960s, of replacing coal by oil, excluding coking coal, the report arrived at a total of 17·5 cents per barrel. On the basis of maximum replacement of uneconomic coal, whilst financing electricity generating stations capable of operating with either coal or oil, in order to provide for reverses of circumstances which are always possible, 'the substitution of adequate security for inadequate security would save Europe 2·6 billion dollars a year'. This would give a sizeable financial saving, even after reckoning the costs of special security stocks of oil and after paying the displaced miners compensation equivalent to their present earnings for the rest of their lives. The purpose of the move would, of course, be to dissuade oil-producing countries from keeping up a boycott for a long period. It was anticipated that in due course disunity among them would occur, as it did in 1956 and again in 1967.

This view of the matter leaves two questions to be answered. The first is political: what government will have the courage to eliminate the coal mining industry abruptly, even if this would be a suitable measure? The second is technical: how long must an oil-consuming country be prepared to hold out without supplies from abroad? A financial aspect is involved in both these questions.

M. Marcellin, French Minister for Industry, estimated in November 1966 [13] that the amount of stockpiled reserves should be equivalent to six months' consumption, which he considered to be long enough to discourage producing countries from an all-out trial of strength. He thought that those states would be unable to bear for more than six months the loss of earnings caused by stopping their exports – unless they were determined on political suicide. Professor Adelman has asserted that some countries might be prepared to maintain a boycott for longer than six months but he believed no united front of the exporting countries could last as long as that without rifts, as experience in 1967 indicated. A series of relaxations in the boycott, achieved through mutual concessions, would suffice to isolate the recalcitrants and re-open trade channels. The saving, by means of rationing, of the equivalent of one month's consumption, together with the two months' stock normally held by the trade and the six months' stockpiled reserves, would enable Europe to hold out for nine months without fresh supplies and it could reasonably be expected that within that time it would be possible to get the boycott lifted.

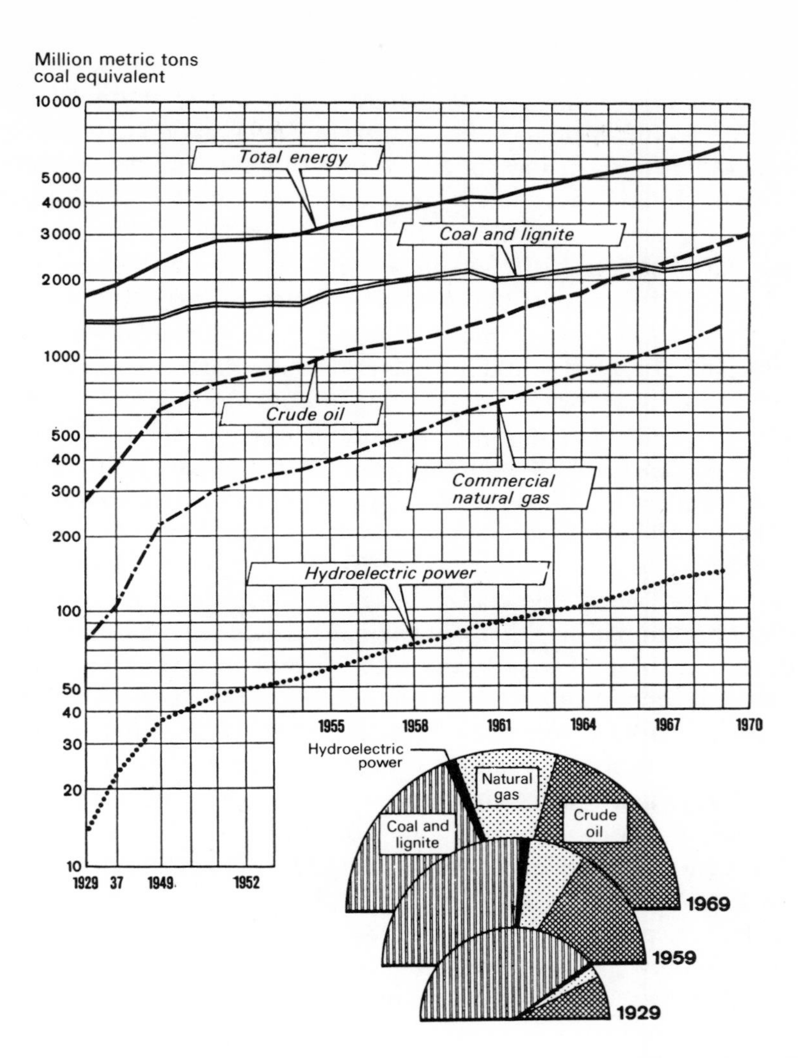

Fig. 11 World energy production and shares of the different forms of energy, 1929–69

Source: Comité professionnel du pétrole

Thus, a stockpiling policy of this kind allows the notion of complete security to be entertained: it permits some certainty of being able, come what may, to withstand the emergency longer than the opposing side. But having worked out the solution it is no use simply leaving the matter there. In order to have that complete protection it must be actually possible to provide the alternative supply if no fresh supplies enter the country for six months: the threat cannot be countered unless the corresponding stockpiling can be achieved. At present the only solution considered possible is underground storage. The French Government has estimated the cost at an investment of one thousand million francs, plus annual costs. It has even been argued that the cost of building these underground reservoirs would only be worthwhile if there could be certainty that a serious crisis would occur every ten or twelve years. But will such crises still be possible if one side holds a decisive advantage over the other? As in weaponry, one side would be thwarted by not catching up with means of reply to bold initiatives by the other side.

Experts can try to work out in terms of coal equivalent the statistical effects of a policy based on seeking almost complete security of oil supplies. But this involves assuming that oil prices are stable or have a downward trend, which was the case during the 1960s but no longer holds good when the market trend has changed to favour the producing countries, as it did in the winter of 1970–71. How is the government to act in a country where the coal industry is a traditional part of national life? There are human problems involving districts with a long history of dependence on the industry. Besides, technical conversion takes time and requires a certain minimum degree of support. At the same time, it seems rather paradoxical to base security on the exclusive use of one material.

In this connection Italy, never having had much coal, is in a way better off than France and Spain, where any progress in the proportional use of oil is made at the expense of the coal mining industry. But is not dependence on a single energy source a cause of weakness?

It must be added that security is not the only aim to be considered. There is the price factor, which must be taken into consideration in any rational oil policy in a major consumer country. In his study, Professor Adelman claims in conclusion that the aims of assured supplies and of cheap supplies are not incompatible. He writes that: 'There is no dilemma of cheap versus secure fuel for Europe, nor for Australia, Japan and other Asiatic nations. The only way to get cheap and secure fuel is to stockpile oil and get rid of coal.' [14] This treatment of the subject seems to be biased in favour of technological considerations at the expense of policy and to be over-optimistic. The argument is put forcefully and the judgment falls like a guillotine. But reconciliation of the two sides of the matter is not as easy as it might seem to be. If all the economic obstacles were removed there would still be psychological difficulties and therefore political obstacles. A point of view does not necessarily gain acceptance because it is rational. In politics allowance must be made for irrational reaction and for the fear aroused when too harsh a truth is revealed. The time factor is needed in order to win over those who have to suffer for the sake of progress.

Even when a political choice had finally been made and had brought into practice the irreversible run-down of coal production and its inevitable replacement by hydrocarbons the active pursuit of security could not be neglected. This must continue at all times, whether the situation is one of war or peace or crisis. The function of politics is to forestall or eliminate possible mishaps, providing a shock-absorber. The aim of mitigating so far as possible any developments harmful to the economy and to national defence remains. This involves using active strategy to combat conflicting strategies, by making use of the two factors of time (stockpiling) and space (diversification of suppliers). Politics is entrusted with the task of correcting trends at home and counteracting hostile behaviour abroad.

The application in practical economics of this political attitude may be expressed in a few simple principles. French planning experts seem to have understood clearly the economic implications of acceptance of the necessity for security of oil supplies. In the proposals stated in the report on the options to be considered in connection with the Sixth Plan (1971–75) they mention that the concept of security of supplies involves 'the physical continuity of supplies essential to the economy; precautions against excessive additional costs (in francs or in foreign currency) in the event of crisis; protection against dominance where this could impose on the national economy a risk simultaneously affecting cost, availability of oil and petroleum products and independence'.[15]

The wording of the statement brings together three factors which should surely be regarded as inseparable in any serious consideration of security of supplies, namely the physical continuity of supplies, precautions against possible crises and the necessity of a certain degree of independence, which must not be too costly nor too isolationist.

There are two concepts of security which should not be confused with each other. One means making sure that there will not be any shortage of supplies and may be called 'all or nothing security'. Another, wider, view is usually represented by the term 'economic security'. For an economy this means being protected against undue fluctuations in the international market or being immune from any pressure that may be exerted by economic or political factors outside the orbit of national sovereignty.[16]

Aim of minimum cost

Countries with large consumption and little production, among which France, Italy and Spain are included, aim at security of supplies but also seek to obtain the oil and petroleum products as cheaply as possible. Therefore in their national policies the aim of minimum cost is pursued simultaneously with that of security of supplies. The latter has been discussed at length; the meaning of minimum cost now needs to be considered. Therefore the implications of the concept of minimum cost will first be discussed, then the available scope for dealing with this matter will be considered and finally the meaning of economic cost will be examined.

The concept of minimum cost

Minimum cost is a reasonable ambition but it can never be fully achieved. Moreover, there is a subtle line of argument linking the notions of cost and security.

Assured supplies and cheap supplies are policies which in practice prove to be difficult to reconcile. This is implied in the saying 'security at all costs'.

Certainly the United States, for example, has had to pay the price of seeking security. The policy there is to keep oil wells in Texas and in Louisiana with very small production in working order and therefore to set the cost price at a very high level: it is much higher than the cost of Middle East oil, even after the expense of shipment from the Persian Gulf to the ports on the Atlantic coast of the United States. That is how the United States has virtually isolated its oil market from the rest of the world. Having attached extreme importance to security, the country is prepared to pay a high price for it. In fact, the price of crude is designed to be such as to render extraction profitable throughout United States territory.

After the Second World War Western Europe was involved in this vast market dominated by the United States. So long as there was a shortage of oil, namely until about 1956, Europe had little alternative to accepting the products and the pricing system of its ally across the Atlantic. It obtained assured supplies but the bill was heavy. The position has since changed with the introduction of some competition. The discovery of new oilfields, in particular those in Algeria and in Libya, gave a sudden increase in supply. The 'independents' which had launched ventures outside the United States but lacked a sales organisation proportionate to their discoveries began to employ price-cutting tactics in order to dispose of the enormous quantities which a combination of skill and luck had given them. The state companies have also tended to affect the international market, upsetting the traditional trading arrangements of the big companies, under which sales are made by the parent company to associated companies. Thus, this sudden increase in supplies and better organisation among buyers resulted in a downturn in the price of crude oil from 1959 to 1969. The drop in price reached as much as 25 per cent, and this without impairing security of supplies. However, since 1970 the international companies have agreed to substantial rises in price precisely in order to avoid a general breakdown in supplies.

Therefore cheapness and assured supplies do not necessarily go together. But the former is a wider concept embracing the latter, since insecurity is a risk measurable in terms of cost. In its report on possibilities for France's Sixth Plan the French Oil Committee, *Comité du Pétrole,* rightly points out that for a consumer country the notion of 'optimum cost' means the cost obtained by considering both security of supplies and the lowest possible price. The report states that:

> The constant objective of policy in oil matters is to achieve assured supplies at the lowest possible relatively stable cost. Of course, it is not possible to succeed fully in obtaining both low cost and security in combination and a balance must be maintained between aiming at reasonably priced liquid fuel and striving for the indispensable security of supply. Hence in determining the optimum cost over a long period the cost of reasonable guarantee against risks of insecurity should be included and quantities which are available only during a short period should not comprise too great a proportion of supplies. This principle also leads to the cost of guarantee of supplies being the lowest possible cost consonant with the aims pursued.[17]

This interpretation by the French experts would seem to be valid for other consumer countries, including Spain and Italy, as well as for France. Like the French, the Spanish and the Italians are seeking a theoretical point of equilibrium between the need for cheap energy and the requirement that there should be no breakdown in supplies, not even for a day. In the long term a price which may be regarded as an 'optimum cost' for the nation involved is decided according to that moving point of equilibrium.

Yet a distinction must be made between the notion of cost and that of price. The price of oil on the American market is high but the overall cost to the American nation is relatively low when allowance is made for the fact that the oil companies of that country are engaged in international operations which contribute to the prosperity of the national economy. Therefore one should not allow oneself to become trapped in the side issue of the price of crude oil. The economy as a whole should be considered before passing judgement. The consideration of fundamental importance is the cost of development to the nation as a whole. It is necessary to weigh up whether or not national assets are created in a process extending from production to distribution, the amount of capital from the country in question invested abroad in oil production, the foreign currency savings achieved, and whether or not reciprocal trade is linked with the import of crude. In sum, before reaching a decision on whether the cost of oil supplies is high or low it is necessary to draw up a full and careful balance sheet of the mutual benefits and disadvantages of the structure of relations set up by trade in oil. Thus, for example, Italy makes up for large imports of oil by re-exporting after processing in its domestic refineries. The amounts of crude entering the country are far in excess of consumption capacity in the Italian market.[18] The flow of imports produces a flow of exports to the countries supplying the crude. Italy exports goods and services with a high market value to the countries from which the crude comes (textiles, metal goods, etc.). Although the bill for energy may be high, the gain by the Italian economy as a whole is reflected elsewhere in the trade balance.

Clearly, in taking stock of a nation's overall economy it is impossible to disregard the cost of other sources of energy when considering the cost of oil.

The need for the national economy to obtain energy at minimum cost is a factor in survival and development in the present conditions of world competition. No policy-maker can disregard those circumstances. In the quest for cheap energy the whole range of sources of production must be taken into account. There one notes the handicaps to France, and to a lesser degree to Spain, arising in connection with the survival of a coal-mining industry inherited from the previous century. In an in-depth study of French energy policy from the Second World War to 1985, entitled *La politique de l'énergie en France de la seconde guerre mondiale à l'horizon 1985,* M. Vilain notes, for example, that 'despite government subsidies of some 3 thousand million francs a year, nationally-produced coal remains dearer than the alternative petroleum products and even proves to be dearer than some of the imported coal. In order to be competitive its price would have to be lowered to 40 francs per metric ton but the price is 60 francs and the cost price is 110 francs.'[19] Accordingly there is a political possibility of playing off oil against nationally-produced coal. For such a measure to be tolerable it would have to be spread over a period of time. But the same author emphasises that 'oil is already under competition from American coal entering our ports and it will in future be subject to competition from natural gas, which is being found in ever-increasing quantities, and later from nuclear energy when that source becomes competitive. It can therefore be envisaged that in due course liquid fuel, which has steadily displaced solid fuel for industrial purposes, will in turn come under competition from both natural gas and nuclear energy.' [20] The large price rises since 1971 have merely speeded up the anticipated course of events. Although oil has been prevailing over coal in Western Europe, it cannot escape competition from other energy sources. Gas is beginning to challenge oil and nuclear energy will indubitably do so in future. In the long term that competition will certainly be a factor helping to curb the rise in oil prices.

For a long time rivalry abroad caused a downward trend in oil prices. However, the results of competition between producers may be offset by another development, namely the risk that the widespread implementation of anti-pollution measures will markedly affect the growth rate of oil consumption, either by slowing down the increase in the use of oil or by causing a rise in the delivered price. Moreover, production is becoming more costly, despite technical advances and greater competitivity. The cost of exploration at sea is between twice and four times that of conventional prospecting on land. [21] But in 1968 off-shore oil accounted for 20 per cent of the total world production of 2 thousand million metric tons. The depth of underwater prospecting has been between 10 and 60 metres but it could in future range between 10 and 300 metres, in which case great technical progress will create a new political problem, because under the 1958 Geneva Convention on the Continental Shelf each coastal state administers the offshore area to the depth of 200 metres 'or, beyond that limit, to where the depth of the superjacent waters admits of the exploitation of the natural resources'. How can international organisations solve the legal situation which will arise? Can they take over this new domain or can they use international conventions to oblige the oil companies to organise collective ownership? Oil has a dynamism of its own which cannot be confined within the narrow limits of national legislation but requires supranational measures. From now on the oil industry will have to contend with special difficulties in that it must make more rapid progress in order to withstand competition from other energy sources, yet such advance cannot be achieved without higher costs.

Possibilities of limiting cost

None of the three consumer countries with which this book is especially concerned can afford to disregard the need to keep the cost as low as possible, whilst preserving reasonable assurance against the risk of insecurity of supplies. But when these general aims are put into effect there are three possible courses – to obtain supplies from international groups, to buy on the 'free market' or to have enterprises with national capital.

If the choice is to be a customer of the large international groups, then it is necessary to be prepared to pay a price in excess of the cost price. The seller expects to derive a substantial profit from its work and the buyer must be willing to provide compensation for the exploration risks which it has not undertaken itself but has left to the other party involved in the transaction. In fact, the vendor disposes of its product at a list price which includes a profit that is, by nature of the factors involved, rather high. After inclusion of the cost of taxes, the FOB price paid by the customer is between 50 per cent and 70 per cent higher than the cost price.

But there is no manipulation by some malicious agency in the way in which ill-informed observers sometimes imagine. Vendors and purchasers perform a sale of perfectly normal type involving two sides with opposing interests. Each buyer acts separately, whilst the sellers are few, well organised, aware of their position and constantly seeking to increase their economic and financial power.[22]

If the buyer dislikes this system, which may appear oppressive, there is another possibility, namely that of making full use of the strength conferred by potential purchasing power and seeking supplies on the 'free market'. There is a free market because some crude oil is outside the integrated networks carrying oil from the well to the pump under the auspices of a single company throughout all the stages of the process. This oil remains outside the traditional networks. Supplies on the free market come from the USSR, the 'independent' producers (most of whom are American) and some international companies (including Gulf, BP and CFP) which have considerable surpluses from the Middle East available. The 'independents' are particularly active in gaining a market in this way. Operating from oilfields in the Middle East and above all in Libya since the mid-1960s, they have large quantities of crude for sale but lack a structure of associated companies linked to a parent company. They are producers only and neglect the possession of 'pumps' in order to concentrate on 'wells'.

On this free market prices were between 20 per cent and 25 per cent lower than through the international networks. Sometimes transactions can be carried out by barter, without the expenditure of currency. But the market was small (5 per cent to 8 per cent of the world market) and the duration of contracts was usually only three to five years. However, so much crude has become available in the last few years that the free market is becoming more important. The quantities being offered are larger and contracts may be for a longer period, for example, ten years.[23] An important effect is that arrangements between the two sides in these transactions produces a spin-off, obliging the international companies in turn to reduce their prices. In times of ample supply the big producing companies with a scale of operations such as that of Shell, Esso, etc. may help the subsidiaries tied to them to meet increasingly keen competition by giving them large rebates. Usually these rebates are kept secret but it is known that they may be as much as 15 per cent or more.

If the two methods described above fail to yield the desired results there is a third possibility. This is the purchase of shares in existing oilfields or prospecting at the risk of the enterprise concerned for oilfields to be exploited by companies operating with national capital. There are in practice three basic stages in this process. First the risks of prospecting and the hopes of satisfactory profits have to be undertaken by either public or private enterprise using national capital. The objective is to hold reserves for the consumer country which can be profitably exploited. The next step is for the enterprise in national ownership to engage in production on its own account, in order to sell in the national currency at a price which includes a profit margin that is not unreasonable, hence not greatly above the cost price. The third phase is reached when it becomes possible to recover the equivalent of local taxation by exporting to the countries in which the oil wells are located sufficient goods and services incorporating advanced technology, hence of high sales value. A bold commercial policy may even yield a positive balance in favour of the country with a shortage of domestically produced oil, or at least an undue outflow of currency may be avoided.

The third method is clearly the best for obtaining supplies at reasonable cost, provided that the stake is adequately distributed between the various producer countries and provided that it is accepted that a policy of long-term investment abroad has to be undertaken and may not yield worthwhile results until many years later. An analysis of this kind leads politicians to consider setting up a state company. They see this as a way of competing with the international majors on their own ground. But this policy produces a heavy financial burden until the state organisation is large enough and has become virtually self-financing.

The concept of cost to the economy

Obtaining supplies cheaply remains a priority aim in the oil policies of consumer countries but it becomes more complicated with increasing diversification of the economy of a country. Rational thinking in this connection involves considerations of overall economic policy. In other words, the concept of cost must be carefully distinguished from that of price. Clearly, the notion of cheap supplies means the cost in relation to the national economy as a whole. Any calculation of that cost must take into account the value of certain items contributing to the final result. It is therefore necessary to think in terms of an overall balance sheet, calculating on the profits side the savings in currency achieved and the goods and services sold and on the other side the oil imports. Only the latter will bear a minus sign.

The concept of cost must be placed in the overall context. As M. Giraud has explained:

> This can mean the price of the transaction, namely the price entered on the invoice sent by the supplier to the purchaser. Alternatively it can mean the cost of oil to the economy, which amounts to the sum that the economy of the purchasing country is to spend in order to obtain a ton of crude oil. One is concerned with the net expenditure, because there may be both expenditure and receipts. But what is important is the difference between the two, representing the cost to the economy.[24]

Earlier in this chapter a distinction was made between two different ideas conveyed by the concept of security, namely absolute security and economic security. Similarly, in discussing the aim of minimum cost one arrives at a differentiation between two views of the matter. The first view is, in fact, more commonplace but it is short-sighted: it goes no further than the object which is immediately apparent, that is the market price. It looks only at the sum on the invoice paid by the buyer and compares that sum with the amount paid by another importer to another supplier. But that method of calculation is too one-sided. It omits to put the transaction into its proper position in the economic life of the country involved. It does not make enough allowance for the security factor which is included. It underestimates the importance of time. A country may have advantageous terms in paying for its oil at a particular time but if a crisis occurs what will happen? Will it still receive supplies? If transactions are resumed after a stoppage what will the price level be when importers have had to yield to pressure from producing countries? By contrast a particular country which usually pays quite a steep price for its oil may have obtained better assurances of the regularity and continuity of its supplies.

The European countries under special consideration, France, Italy and

Spain, try to combine the three possibilities for a minimum cost policy. But in order to judge the success or deficiency of the procedures adopted by them one must look beyond the price of the transaction. The most important feature is the 'net expenditure' made upon purchase of a ton of crude. Outgoings may be more than compensated by what is received in return. Imports must be viewed in the light of the exports which they render possible. Accordingly, it may be necessary to spend large sums on the operations of state companies but that total must be weighed against the benefits obtained by the country involved as a result of the operations. Conversely, one must consider what the price would have been if the intermediary had been a foreign company. Finally, one must assess what form of organisation is capable in the long term of providing the best protection at the lowest cost against risks of insecurity of supplies. Is it a state company, which necessarily adopts a more political attitude? Is it a private company adopting conventional economic tactics? The former may emerge the winner in struggles between opposing national ambitions but it lacks the flexibility of the latter when there is a sharp rise in political tension between two countries. Moreover, because of its size and its international scope a private company may achieve better distribution of its involvement and may possess a more effective range of possibilities of dealing with a crisis because its operations are spread throughout the world.

A further question is that of which type of enterprise is more concerned to set its prices low. What line do oil companies take vis-à-vis the host countries? For the sake of harmony the two sides tend to agree readily to a price increase, provided that it applies to all concerned. A state company cannot succeed without joining in that system. But given that situation, which kind of organisation protects the interests of the multitude of consumers, each purchasing separately and to whom a low price is important? A state company is initially set up to assist these customers but there is a risk that the process of its development will cause it, in turn, to adopt attitudes of a dominant seller imposing its prices on buyers lacking collective powers. The State then acts as the accomplice of the company in any decision between protecting the interests of the country's consumers and the desire to obtain the success of the state enterprise constituting the essential means of pursuing an oil policy aiming at national independence.

For the sake of a certain degree of security at national level, the ambition of obtaining oil cheaply ceases to be an absolute necessity. The main objective becomes that of 'security at all costs'. Thus the aim of minimum cost becomes subordinate to that of security, whereas for a healthy economy well-designed for international trade the opposite policy should be practised.

To summarise the essence of this chapter, France, Italy and Spain, as large consumers and small producers of oil in their own territory, practise the twofold policy of seeking assured supplies and endeavouring to obtain oil cheaply. These two objectives are pursued simultaneously and there is difficulty in reconciling them completely: the relationship between them is dialectic from the point of view that security has to be paid for and its cost can be calculated in cash terms. Therefore, transcending the two aims that have been discussed there is an overriding desire to achieve an 'optimum cost' as shown by the result of several sets of calculations, namely adequate return for production factors, payment of fiscal charges, the net balance of assets and capital involved in the processes arising from the necessity of importing oil, and the measurable cost of a reasonable and stable guarantee against the overall risk of insecurity.

Notes

[1] It has been estimated that USSR consumption of oil and petroleum products would be multiplied by a factor of 3·2 from 1967 to 1980, which would give a level of 595 million metric tons at the end of that period. As regards oil production in the Soviet Union, some experts give estimates of between 600 and 620 million metric tons in 1980. This target means that output would increase by about 85 per cent from 1969 to 1980, which has been regarded as ambitious, since amounts discovered up to 1969 had been insufficient for that level of production. Cf. *Petroleum Press Service,* January 1970, p. 4.

[2] *Petroleum Press Service,* March 1970, p. 82.

[3] P. Guillaumat, 'Le pétrole dans la défense et l'économie nationales', *Revue de défense nationale,* January 1968, pp. 5–17.

[4] H. Longhurst, *Adventure in oil,* William Clowes and Sons, London 1959.

[5] J. Baumier, *Les maîtres du pétrole,* Julliard, Paris 1969, pp. 65–6.

[6] Some thirty years earlier Clemenceau himself had opposed a proposal in the French Senate that protection should be given to national oil refining. He then said, 'When I need oil I simply go along to my grocer'. This shows how attitudes are changed under the pressure of the needs of wartime economy. Cf. J. Masseron, *L'économie des hydrocarbures,* op.cit., p. 269.

[7] On this point see the book by P. Clair, *L'indépendance pétrolière de la France, I, Le théâtre de guerre,* Ed. Cujas, Paris 1969.

[8] Memorandum from the EEC Commission to the Council, dated 16 February 1966.

[9] *Official Journal of the European Communities,* 23 December 1968.

[10] D. Murat, *L'intervention de l'Etat dans le secteur pétrolier en France,* Ed. Technip, Paris 1969, p. 210.

[11] *Petroleum Press Service,* March 1970, p. 83.

[12] M.A. Adelman, *Security of Eastern Hemisphere Fuel Supply,* Department of Economics, Massachusetts Institute of Technology, no. 6, December 1967.

[13] *Journal Officiel* (France), National Assembly debates, November 1966, p. 4321.

[14] M.A. Adelman, op.cit.

[15] Commission de l'Energie, Comité du Pétrole (France), *Propositions pour le rapport sur les options pétrole,* 4 February 1970. Roneotyped document from the Commission regarding the Sixth Plan, p. 2.

[16] Unpublished address by M. Giraud, Director of Fuel in the French Ministry of Industry, to the UN Economic and Social Council. Supplement to the record of the session on Thursday, 13 June 1968.

[17] Commission de l'Energie, Comité du Pétrole, op.cit., p. 2.

[18] In 1969, for example, Italy headed the EEC countries in imports of crude: 102 million metric tons (West Germany 89 million metric tons, France 88,400,000 metric tons). Final consumption of crude in Italy was only 53,300,000 metric tons. Cf. *Bulletin de l'industrie pétrolière (BIP),* 1536, 3 March 1970.

[19] M. Vilain, *La politique de l'énergie en France de la seconde guerre mondiale à l'horizon 1985,* Ed. Cujas, Paris 1969, p. 348.

[20] Ibid., p. 313.

[21] J. Baumier, *Les maîtres du pétrole,* op.cit., p. 243.

[22] P. Desprairies, Public Relations Officer of ERAP, *La politique pétrolière française,* talk given to the Inspection des Finances (French Inland Revenue Inspectorate), 25 April 1966.

[23] An illustration of this change is provided by the ELF-Occidental contract signed in 1968 concerning Libyan crude.

[24] Giraud, Internal document of UN Economic and Social Council, op.cit., p. 6, note 15.

2 Ambitions of the Producer Countries

Whilst oil affairs are basically three-sided, involving producer countries, private or state oil companies and consumer countries, so far the only aims examined in depth in this book have been those of the consumer countries. However, in pointing out the characteristics of Spain, France and Italy among Western Mediterranean coastal states providing the geographical frame of reference for this book there has been mention of the special relationships between those three countries and the states on the oil-producing North African side of the sea, namely Morocco, Tunisia, Algeria and Libya. Morocco and Tunisia are only marginally involved, being both small consumers and small producers of oil. At present oil discoveries in Tunisia have been more promising than in Morocco. On the other hand, Algeria and Libya do serve as representative examples, being genuine counterparts of the three European countries. As net producers, being small consumers and large producers, these new members of the oil club appear in the market as sellers. Libya and Algeria are members of OPEC[1] and OAPEC[2], which serve as think tanks and trade associations for protection of the interests of the oil-producing countries.

Whatever divergences there may be in specific matters, the objectives of Algeria and Libya in oil affairs are broadly similar in general intent. These objectives will now be examined under three heads, as it is clear that in connection with the oil which is predominant in their national economies these two countries pursue, sometimes successively and sometimes simultaneously, three main objectives – a fiscal objective, an objective of exercising control of oil production in their territory and an objective of national development.

Fiscal objective

Except in the cases of the United States and the USSR, the great majority of oilfields have been found in economically underdeveloped territories, therefore initial exploitation of oil resources has not usually been undertaken by the producer country itself. Foreign companies have

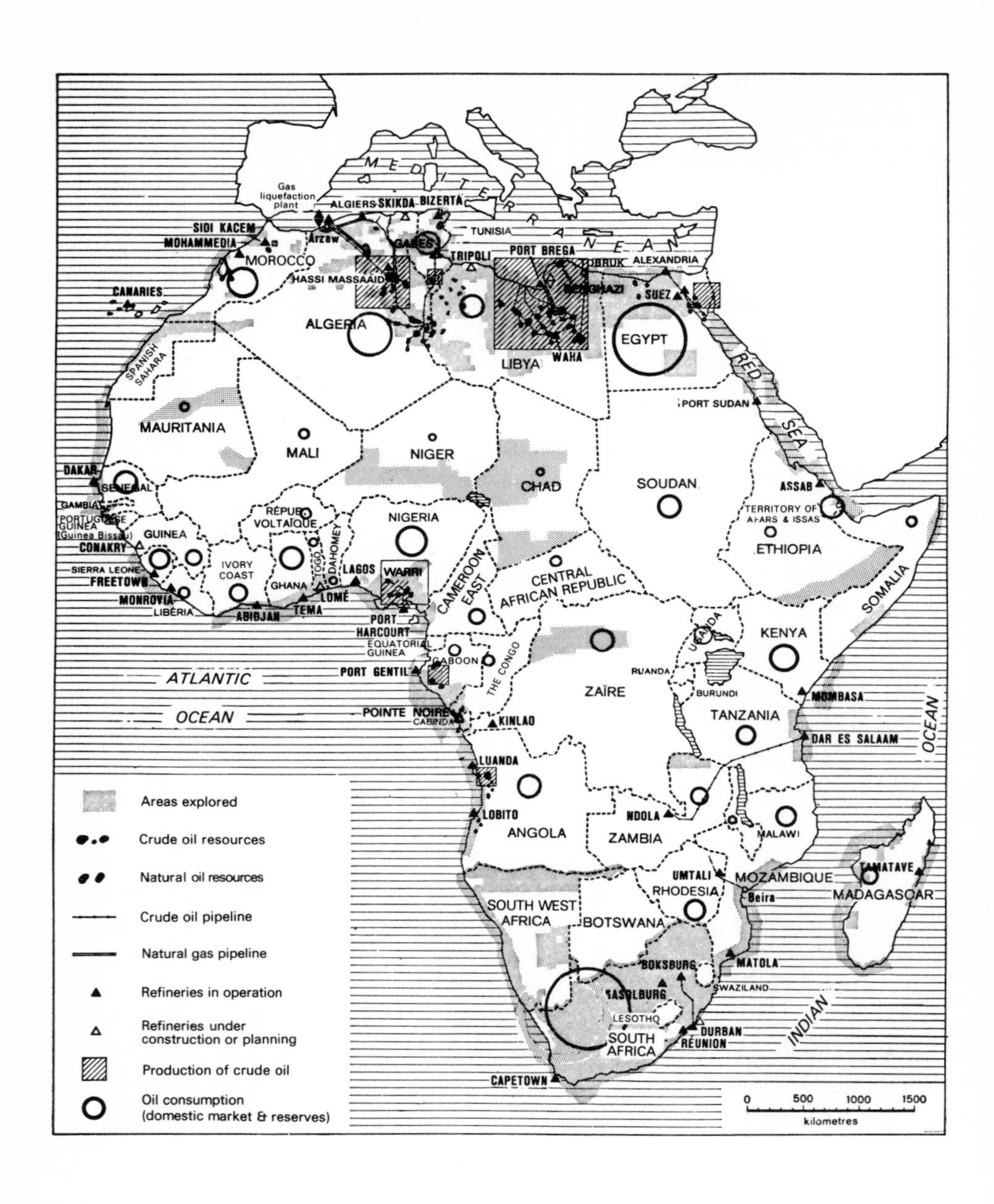

Fig. 12 Oil in Africa (situation at the beginning of 1974)

Source: Comité professionnel du pétrole

carried out the exploration and obtained exploitation concessions, because the host country is financially too poor or has too few skilled personnel to venture into these operations demanding a high level of expertise and heavy investments. However, the producer governments definitely expect a share of the benefits from the exploitation of the riches of their subsoil. Chronologically, the first objective to receive attention from a producer country is in the fiscal sphere. Having regard to the fact that the companies are the means whereby profits are derived from the resources, the producer governments devote their attention to the companies' income from their operations rather than the capital involved. Asserting their sovereignty and at the same time collecting large amounts of national revenue, the oil states impose heavy taxation on the proceeds of the companies' operations. As in any well-designed tax policy, the aim is to obtain a proportion of the gains without making such a large hole in the funds that investment and expansion are discouraged and hence in due course there is a reduction in the taxable base. The principle which must be applied is the normal one of taxing the gains from an activity without restricting the activity itself, in order to avoid killing the goose which lays the golden eggs.

The general trend is towards increasing the revenue obtained by the state, whether by higher taxation of overall profits, by various levies, through reference prices or by adjusting allowances against tax.

Taxation of profits

There is a great deal of argument between producer countries and the big companies about taxable profits. The producer governments tend to enlarge the scope, claiming that the profits affected should include all stages of handling the products, from the well as far as the dock – some even say as far as the pump in a foreign country. There is also controversy as to whether the profits should be taxed before or after other taxes have been applied. Should royalties be treated as a production cost not deductible from the taxable profit or as an advance payment to the state against its share of the profit from the whole sequence of operations? Algeria and Libya have favoured opposing alternatives according to developments and they still differ from each other quite considerably in their attitudes in the matter.

When Algeria was under French administration its oil legislation imposed direct taxation on the gross profit at the rate of 50 per cent. However, the mining royalty was treated as an advance instalment towards the 50 per cent. This royalty was 12·5 per cent for liquid hydrocarbons and 5 per cent for hydrocarbons in the form of gas. Its payment was obligatory in all circumstances, whatever the financial results of the operations in progress, in order to supply funds to the public enterprises.

These principles were by and large perpetuated in the 1965 agreements between France and Algeria. For companies directly affected by the arrangements, essentially companies with French capital, the royalties of 12·5 per cent were deductible in calculating direct taxation but were retained by the Treasury if there were no profits. However, those companies were subjected to progressive increase in the rate of the direct tax, which rose to 53 per cent for the next three years, then 54 per cent for 1968 and 55 per cent for 1969 and 1970. But the whole of this rationally organised system was swept away in favour of semi-nationalisation of the French companies in 1971.

In Libya a somewhat different procedure has been adopted, although the objective has been the same. There the royalty has not been deductible before direct taxation. As a result of that relatively new state's membership of OPEC, the ten-year-old oil legislation was amended in 1965 in order to align the regulations laid down by the Government of the country (at that time still a kingdom) with the rules determined by the 1964 OPEC Conference in Djakarta. From then on the Libyan Government adopted as the basis for calculating royalties and tax not actual prices but posted prices. Royalties were set at 12·5 per cent of posted prices and the 50 per cent share of profits payable to the state was also based on posted prices. However, a small tax allowance could be claimed for distribution and a 6·5 per cent rebate provided for under the OPEC formula might be obtainable. Under the Franco-Libyan agreement of April 1968 setting up the state company Lipetco and placing it in the public sector in Libya, the increase in taxation was carried still further, as the agreement provided for the royalties not deductible before profits tax to be raised from 12·5 per cent to 15 per cent as the output of the joint enterprise grew.

Whatever the details of the measures adopted, from 1960 to 1970 the producer countries have consistently sought to obtain increase in the share of profits payable to the state. Their successful actions in 1971 and 1973 have accelerated the trend. The rate of 55 per cent introduced in 1971 remains in effect now, in 1974, but the OPEC experts propose raising it to 87 per cent, with the royalty remaining at 12·5 per cent of the posted price. The intention is to limit the companies' profit per barrel to 50 cents.

Rules for calculating tax allowances

French legislation applicable in Algeria had provided for the possibility of deducting from taxation for five years the sums spent on developing oil wells. This arrangement was modelled on customary practice under United States legislation, which seeks to encourage the companies to carry on continuous exploration and gives them the means of doing so. The Franco-Algerian agreements of 1965 removed this provision and imposed more conventional rules for calculating allowances. The purpose was to limit the scope of the activities of private enterprise, making greater medium-term demands for revenue for the national economy.

In Libya the Government took the same line, although it was less concerned with the legislative basis for its actions. In fact, there was a regulation permitting a 25 per cent deduction before direct taxation. But the government put pressure on the companies to renounce this allowance 'voluntarily' as soon as they found oil. Because pressure of this kind was accompanied by threats of positive action, most of the companies soon decided that it was better to avoid reprisals by giving way regarding the facilities for developing wells.

Thus, either by the application of legislation or by carefully applied pressure the two governments have arrived at one and the same endeavour to achieve a large increase in the amounts of revenue flowing into state funds as early as possible.

The base for taxation

There has been bitter dispute between the companies and producer governments regarding the exact basis for calculating profits tax. Should it be the prices actually charged? If so, there has been a downward trend because of increased supply and the discounts customarily granted by sellers. Should tax be based instead on posted prices or on listed prices? These have the advantage of being very stable, of varying very little because of temporary fluctuations and of being more even between producer countries, taking into account the distance of the oilfields and the ports of shipment from the major consumer centres. In 1959 and again in 1960 the listed prices decided by the oil companies were lowered. Naturally, the producer countries protested against these reductions. The 1960 drop in prices was indeed an immediate cause of the formation of OPEC as an association for the protection of the interests of the producer countries. From the outset one of the foremost aims of that collective body has been to prevent the catastrophic effects on the flow of revenue of a drop in prices.[3] At the Djakarta Conference in 1964 agreement was reached on a compromise whereby the base for calculating profits tax should not be actual prices but listed prices less certain percentages.[4] Thus, one of the main points in the demands of the producer countries has become adherence to listed prices, or even assurance of their consistent rise, whatever the actual movement of prices. Moreover, the producer governments want to obtain control of the determination of listed prices, by gradually stripping the companies of their former prerogatives.

Under French administration the oil prices in Algeria were governed by Article 33 of the Petroleum Code, whereby selling prices were aligned with current prices in the international market. There were two reasons for this arrangement, firstly to prevent parent companies from giving their subsidiaries special prices, since this practice was detrimental to the sums received by the national authorities, and secondly the positive purpose of enabling the companies operating in the Sahara, which were recently formed and still financially vulnerable, to obtain substantial profits. The avowed intention was to save the newly formed companies from being unable to compete with powerful private enterprise groups possessing funds, in most cases capitalist, out of all proportion to theirs.

The 1965 agreements imposed a fairly high tax reference price on the foreign companies operating in Algeria, namely 2·30 dollars per barrel FOB from Skhirra, 2·35 dollars FOB from Bougie and 2·365 dollars FOB from Arzew. By contrast the bases set for the companies with French capital gave an allowance against the listed prices and were from 2·04 to 2·09 dollars per barrel. In return for this discriminatory measure France undertook to assist in the industrialisation of Algeria.

After the coup d'état of 1 September 1969, placing Libya among the 'progressive' Arab régimes, the tone of the spokesmen of the leading African oil-producing country became much more aggressive. In 1970 Libya demanded considerably more than the 10 per cent rise in posted prices which governments under the monarchy had sought. It has continued to make vigorous protests about price levels, alleging that these are too low by comparison with the Middle East countries.

In the case for higher prices the same arguments have been heard from Tripoli as from Algiers, namely that the selling price of Libyan crude is unreasonably low in proportion to its exceptionally high gravity, its low sulphur content and its proximity to the European markets. Despite the threat made by the Prime Minister, Colonel al-Qadhafi, at his surprise appearance at the first meeting of the new régime with the representatives of the oil companies, that 'people who have lived for five hundred years without oil can manage without it several years longer in order to obtain their rights', none of the groupings involved can afford to make a wrong move. The dispute about posted prices is not insoluble.

Both producers and consumers admit that these prices are justified for two reasons. The first reason is the negative one that there is no other means of valuing the prices of crude. Except in long-term agreements with widely known stipulations, sales are made within networks of members of the same group or under agreements relating to a short period. There are no exchange organisations for oil, so that secrecy and ignorance of current actual prices prevail. Posted prices have the merit of acting as indicators. The other reason, which is positive, is that setting official prices gives the various participants in oil affairs a rough idea of the values of each of the various crudes: it gives the consumer countries a kind of skeleton plan from which to draw up their contracts; it also has the valuable function of providing a stable base for the calculation of royalties and tax.[5] However, the system has the disadvantage of artificiality. There is never a downturn. Increasing deviation from the true price of transactions is obtained.

Limitations on tax policies

Taxation on oil by producer countries therefore shows a continuing rise and sudden peaks. The producer governments try to make use of all the means simultaneously: they raise bonuses and royalties and multiply charges and premiums, either tacitly or officially; they increase pressure on profits; they demand a stable guarantee at the level of posted prices when the actual prices of transactions adopt a downward trend.

This is understandable from the political point of view because for Algeria oil is the most important item in the national economy, representing nearly three-quarters of income from exports. For 1968 the total oil revenue of Algeria, including indirect receipts from oil, was 1,115 million dinars.[6] Two years earlier the figure was only 500 million and the mean growth rate of production from 1962 to 1967 was only 13·5 per cent.[7] As in the case of all the countries with sizeable oil production, these overall figures reveal an increasing discrepancy between the production growth curve and the curve for increase in revenue.[8] In Libya the military revolution of 1 September 1969 did not cause any sudden transformation in oil statistics but merely meant stronger government pressure on the foreign companies. In 1968 the state's receipts from oil were already nearly 1,000 million dollars.[9] In 1969 Libya's revenue per barrel had reached a higher level than anywhere else in the Middle East.[10]

Yet there are limitations on this continual increase in the tax burden. If the drain on their funds is too great, then the companies' ability to provide investment from their own funds will be jeopardised – and investments in this particular industry are necessarily on a large scale. It is difficult to gauge the position in that connection, because the profits of the companies are always shrouded in mystery.[11] This is confirmed by the fact that in practice the oil companies are prepared to let the local authorities have large sums as 'sweeteners', if it is found necessary to do so as the price of avoiding constant argument about their activities. The host governments regard a steady rise in receipts of that nature as a reliable guarantee of the level of their revenue; this is also a form of insurance against risks in a market subject to sharp fluctuations in supply and demand. But a consideration which P. Clair omits to mention is that of competition at global level between oil companies obtaining widely differing profits according to the region of the world where they operate and the effects of national legislation. Excessive tax demands in those countries where the greater part of the operations of the European state companies are based may curb their growth and actually strengthen the 'majors', who are still in a position to make substantial gains elsewhere in the world.[12] Such a course conflicts somewhat with the proclaimed aspirations of ousting from their territory the big 'monopolistic and imperialist' companies. In fact, the producer countries are preoccupied with a different problem. It seems that the host countries are trying to exert all the more pressure because they envisage the decline of the oil industry sooner or later. In addition they are wondering whether Europe will always depend on the Middle East and North Africa for its oil supplies.

Therefore they are endeavouring to obtain the maximum revenue from oil in the short space of time left.

This explains the leaps in the oil revenues of the producer countries after the 1971 and 1973 crises, as the table on the next page shows.

However, the eventual risks of this systematic escalation are beginning to be perceived. If oil prices rise too quickly the producer countries are bound to come into open conflict with other producing states which are less demanding as regards rates of levy but catch up through the total quantities produced. If the members of the producers' group should become too greedy individually, then their collective organisation could in time break up. Moreover, the consumer countries which should be potential allies in this struggle against the organised strength of the companies may not be prepared to bear this constant rise in costs. I have stated in the previous chapter that the desire to obtain oil cheaply is one of their main aims. Some observers even point out that the trend towards a worldwide rise in taxation may perhaps be followed by competition in lowering taxes. If many new oil-producing areas were to be discovered, then it does seem possible that the producer states might seek to attract companies by offering them concessions bearing low rates of taxation.[13] But by that time the whole situation may have changed,

Oil revenues of the producer governments
(in million dollars)

	1965	1970	1972	1973[a]	1974[a]
Saudi Arabia	655	1,200	3,107	4,900	19,400
Iran	522	1,136	2,380	3,900	14,900
Venezuela	1,135	1,406	1,948	2,800	10,000
Kuwait	671	895	1,657	2,100	7,900
Libya	371	1,295	1,598	2,200	8,000
Nigeria	nd	411	1,174	2,000	7,000
Iraq	375	521	575	1,500	5,900
Abu Dhabi	33	233	551	1,000	4,800
Algeria	nd	325	700	1,000	3,700
Other countries[b]	16	150	222	550	1,700
Indonesia	nd	239	555	800	2,100
Qatar	69	122	255	400	1,200
Total	3,847	7,933	14,722	23,150	86,600
Mean revenue per barrel exported (in dollars)					
	0·77	0·92	1·47	2·05	7·69

[a] World Bank estimates
[b] Excluding North America and the Communist bloc

Source: *The Petroleum Economist,* May 1974, p. 165

because some countries may have decided to take their oil affairs into their own hands.

Objective of exercising control of national oil production

The fiscal objective can be merciless. The various levies on the income of companies might be so heavy as to jeopardise even the capital. But to go to such extremes would create a danger of discouraging the operations of the companies and thus destroying the sources from which regular receipts are derived. For that reason neither side goes beyond certain limits, because however bitter their feud about taxation may be, it is not in the interests of either to ruin the other. The fiscal objective does not endanger the existence of the companies and the companies consent to paying a high price in order to be allowed to carry out their own business policies and take precautions to protect themselves against arbitrary tax demands. This is the only explanation for the connivance in a climate of hostility that is practised between consumer countries, oil companies and host countries in the Western Mediterranean coastal states.

The same ambiguities occur when the producer states propose to pursue an objective of taking over control of oil produced in their territory. The fact is that no state in the Near East or the Middle East initiated the discovery or the exploitation of the oil hidden in its subsoil. Foreign companies were encouraged by foreign governments and used their own personnel and their own capital to make local resources profitable. Awareness of this original shortcoming weighs heavily among the present feelings of resentment. Urged on by a mood in public opinion which is quick to take umbrage at the operations of these aliens, the producer governments have in the course of time become all the more sensitive about their own tardiness and aware of the possibilities open to them in world oil strategy in order to remedy their initial failings.

In the African countries of the Mediterranean basin nationalism in oil affairs is becoming the dominant ideology. This lends itself to combination with a 'revolutionary' fervour directed towards removing all vestiges of the former subservience. It assumes a socialist colour in order to be better designed to signify collective rights to interests accused of the three vices of being foreign, capitalist and private. With varying degrees of success and skill the governments of the host countries set out to obtain control of the oil extracted from their territory. But here, again, the advance of the process may vary according to local circumstances and political opportunities. There are subtle distinctions enabling the two extremes of partial control and complete takeover to be included under the same label: the one might be termed participation and the other nationalisation.[14]

In order to examine the situation arising from this objective in the producer countries of North Africa let us consider in turn partial controls imposed on foreign countries, next the development of a national sector of the oil industry and then the prospects of complete nationalisation.

Partial controls imposed on foreign countries

The producer countries have many means at their disposal for directing and controlling the activities of the foreign companies operating in their territories. Official regulations may be adopted, either in relation to exploration only or affecting the whole sequence from taking the oil out of the well as far as shipment, by following the successive processes applied to the actual product.

Libya has always taken care to see that the prospecting zones are fragmented from the outset, so that the companies concerned are obliged to cover the immense area involved. The average size of concessions in this enormous semi-desert territory is between 3,000 and 4,000 square kilometres. The door was opened wide to the majority of the big companies but also to many small 'independents'. In order to force the entrepreneurs to press ahead quickly with their explorations, the law of April 1955 stipulated that applicants must pay the greater part of the charges for their prospecting licences in three stages, at 5, 8 and 10 years, in the event of unsuccessful exploration. This division of the territory into small segments gave Libya the benefit of the endeavours of groups working on neighbouring sites to match each other's progress. The fragmentation makes the oil map of Libya look like a very complicated jigsaw puzzle. In another measure arising from anxiety for rapid results, the law also required the holders of exploration licences to begin their work by a specified deadline and to spend a previously approved minimum amount on operations on the site.

In the event the 'independents' and the big international companies soon discovered prodigiously rich reservoirs. Now Esso, Texaco and Marathon provide the bulk of Libyan production. In 1968, for example, Esso supplied 29 per cent (24 per cent in 1969) of the overall output, making payments totalling some 300 million dollars to the Libyan Government.[15] It was a well-known fact that during the time of the kingdom the authorities were unable to make these mammoth companies comply with the official prices which Libya tried to enforce. The more power to control income in its territory the government seemed to gain, the more it became overwhelmed by events; it gave the impression of lacking the human and technical resources for imposing its will on opposing interests which acted openly but were well aware of their strength and their rights. The greater the success of the overall oil policy became, the more difficult it proved to be to achieve the aim of controlling the large companies. In Algeria, by contrast, the government which emerged from the war of independence found itself in a situation which it could not manipulate to suit itself. It was unable to choose the companies with which to deal in oil affairs as Libya had done. The allocations and the relative positions in the oil industry had already been arranged by the French. That is why the Algerians have tended to make the heaviest demands on the most powerful concerns, seeking to bring the firmly established companies under pressure. The requirements imposed have become more severe in the course of time, involving control of movement of capital, the obligation to algerianise the personnel of enterprises, formation of registered companies with headquarters in the national territory, etc. Algerian leaders have frequently condemned the companies' practice of transferring profits and capital to other countries. At regular intervals they have denounced the economic and financial 'insularity' of these companies in keeping locally only the amount of capital strictly necessary to carry on their local operations and sending the rest abroad. For that reason the Algerian Ministry responsible for the national economy notified the oil companies of its intention that henceforth 50 per cent of their turnover should be kept in Algeria; it was planned to raise this figure later to 66 per cent.[16]

In the case of the co-operative association between France and Algeria, nicknamed Ascoop, a different system was applicable: the French companies could use 40 per cent of the income in France but the percentage was to be reduced until it remained at 25 per cent from 1972 onwards.[17]

The motives of the authorities responsible for this policy of control of the funds of the companies are fairly easy to understand. The intention is to oblige the entrepreneurs to keep much-needed currency in the country; the country is therefore enabled to use capital deposited in the local banks for national purposes; the oil groups are obliged to make use of goods and services available from Algerian sources instead of placing orders abroad in these matters. But this strictness is precisely what the companies dislike, in that oil operations on an international scale demand great mobility of capital. The funds gained locally cannot be used to finance prospecting and exploitation programmes elsewhere, and such activities are a more attractive proposition than the anticipated future of the oil industry in any one country. Moreover, oil operations demand high quality equipment and services which cannot always be obtained on the local market.

In more general ways the foreign companies are constantly subjected to a certain distrust. This is the explanation for the multitude of controls which producer governments impose on them and the uncertain effect of the contracts which they are permitted. The wording of a recent OPEC Resolution reflects clearly this general attitude of suspicion, forcefully asserting the need for rigorous controls and constant reconsideration of the provisions of contracts:

> When a Member Government is not capable of developing its hydrocarbon resources directly, it may enter into contracts of

> various types, to be defined in its legislation but subject to the present principles, with outside operators for a reasonable remuneration, taking into account the degree of risk involved. Under such an arrangement, the Government shall seek to retain the greatest measure possible of participation in and control over all aspects of operations.[18]

In other words, contracts are to be reviewed frequently, in the light of circumstances. But change in the situation often occurs in oil affairs. And it is a risky matter to set up a company when the government of the host country seeks to be the sole arbiter of the way in which that company is to develop. If a company refuses to accept unilateral decisions it may be penalised by the formation of a competing company belonging to the nation concerned or by expropriation pure and simple.

Development of a national oil industry

However punctiliously foreign companies are controlled, this is only the first stage in the procedure adopted by a state wishing to assert its sovereignty over the oil extracted from its subsoil. A more decisive move is made where the state itself becomes an entrepreneur, sets up the general apparatus and uses this to work its way into the system devised by foreigners. By this means governments endeavour through the act of creating a framework to overcome the 'inherited backwardness' which exists in emergent nations because of their poor knowledge of oil technology and of the workings of the world market for hydrocarbons. This procedure is even given priority in the OPEC Resolution of 1968 mentioned above. Contracts with foreign companies are envisaged only if a Member Government is unable to develop its hydrocarbon resources directly. The Resolution begins by urging that 'Member Governments shall endeavour, so far as feasible, to explore for and develop their hydrocarbon resources directly'.[19]

In 1963 Algeria formed a state company, Société nationale pour le transport et la commercialisation des hydrocarbures, which is commonly known as Sonatrach. Libya took a corresponding step later; when the coup d'état occurred in 1969 its company had already been in existence for a year. This was the Libyan Petroleum Corporation, known as Lipetco, which had been formed in collaboration with ERAP/SNPA under an agreement signed in April 1968. Since the military régime assumed control the powers of Lipetco have been strengthened. In order to emphasise more clearly the public nature of the enterprise the country's leaders renamed the company Linoco, or the Libyan National Oil Corporation.

Now it is referred to as Noc, or the National Oil Corporation.

In spring 1970 Algeria and Libya even decided to create a joint company to undertake exploration and development operations in the territories of both countries. The new company is to be a joint venture of Sonatrach, which is already firmly established, and Noc, which was still in its infancy, having been created five years later than its neighbour. This initiative may be explained by political, economic and financial reasons. The two governments have fairly similar ideological attitudes and they wish to give each other mutual support in order to avoid allowing private and foreign enterprise to enjoy a favoured position in developing the oil resources of the two countries. For operations of this kind Libya can supply the large quantities of capital which it has been gathering in but it is woefully deficient in technical equipment and in the supply of skilled labour. Algeria has the complementary strengths and weaknesses. Its sources of finance are limited but it is beginning to acquire skilled workers and suitable equipment for geophysical survey and drilling. Sonatrach could be the active partner in the practical operations of enterprise, with Noc as the main supplier of funds. [20] The plans are still only in the outline stage but they are indicative of the oil ambitions shared by Algiers and Tripoli.

The official communiqué issued at the end of Colonel al-Qadhafi's visit to Algiers on 19 April 1970 stated that:

> The two sides express their approval of the agreements and co-operation arrangements concluded in respect of all stages of the exploitation of oil resources and their integration into the national economies of the two states in order to accelerate their development and eliminate the backwardness inherited from imperialism, a situation which has created a united front between the two states against all the forces of oppression, monopoly and exploitation.[21]

Clearly the Algerian leaders are the moving spirits in the creation of this venture. The basic principles for their oil policy have been made clear often enough in international forums. The Algerian Minister of Industry and Energy expounded them as long ago as 1967, at the Sixth Arab Petroleum Congress. The philosopher and orchestral conductor in Algerian oil policy made it clear that the goal was to assume the position of privileged counterpart of the consumer countries by creating a large public sector. He saw this as a means of completely short-circuiting the international companies. In fact, he told the delegates of the Arab countries that the aims pursued by Algeria were causing it to promote 'a new policy, that of preparing the objective conditions for extricating ourselves from the situation in which we find ourselves at present, a situation marked by domination from the cartel of the large foreign companies in oil affairs. These objective conditions consist of creating a national oil industry capable of forming in the producer countries the basis for a policy centred on direct negotiation with the consumer countries'.[22]

In the event this 'new policy', causing the state to become an entrepreneur, is encountering a great deal of difficulty at the practical level. However, the Arab countries, among which Algeria and Libya are included, are engaged in improving their understanding of the problem and have already worked out a rationally-constructed doctrine.

The theme of state companies has been prominent for several years in the discussions at Arab petroleum congresses.

Did not the Sixth Arab Petroleum Congress, in March 1967, recommend the strengthening of national oil companies and call for the establishment of similar companies and organisations in Arab countries where none existed at that time?[23]

The Seventh Arab Petroleum Congress, held in Kuwait from 16 to 22 March 1970, went beyond these general propositions and adopted a technical approach, seeking to make a serious study of the advantages and limitations connected with a national public sector, before recommending its creation as a matter of priority. The first recommendation emphasised the assistance which the Arab states should give to their national companies and the need for these companies to work together. A paper on 'The role of governments in supporting national petroleum companies' was given by Dr Hussein Abdullah, Professor of Petroleum Economics at the University of Kuwait, and Ahmed Sayed Omar, Chairman of the Kuwait National Petroleum Company (KNPC).[24] The authors declared that the main aims of these companies should be 'to take over local distribution, to develop national oil resources, and to enter into partnership with, diversify and supervise foreign concessionaires'.

These aims might lead a national company to engage in production by itself or in partnership, to set up drilling subsidiaries and so forth. They might also be involved in transport and in owning or participating in refineries.

Some governments gave more help than others. Government backing was needed through direct financing or loan guarantees, permission to build up reserves and to employ foreign earnings overseas, favourable treatment in taxation, the provision of services, and the award of contracts. In the authors' view the Government should negotiate for crude oil and gas to be made available by foreign concessionaires and should try to conclude agreements for joint projects with companies of the consuming countries. Speaking as economists, the authors pleaded for the national companies to be kept free of red tape and given flexibility.[25]

However, it is important not to overlook the present shortcomings of these state companies when they try to engage in activities abroad. In this connection the dilemma facing them is manifest. Either the scope of their operations abroad is too limited and too insignificant or they are obliged to go into partnership with foreign associates in order to gain a foothold in the international market, which means forfeiting independence of action. Even if these companies join forces their sales effort is weak because of the limited range of their crudes and the difficulties of their financial position. If they venture into the general area of world competition they are confronted by the power of the majors, the independents and the governments of the consumer countries. The most promising possibility is to work with the state companies of the consumer countries where those companies have an international scale of operations. But that is only possible if genuine efforts are made to find fair solutions to the initial divergences between the interests of the two sides. If they quarrel the reaction may be complete nationalisation of foreign oil companies operating in the producer country.

Prospects of complete nationalisation

The presence of foreign entrepreneurs has always been tolerated rather than accepted by the Arab countries. The governments are strongly tempted to make nationalisation of the foreign companies a priority aim.

The Kuwait daily newspaper *Al Ra'i El Ame* reviewed this important matter at the beginning of 1970. Dr Mossadegh's disastrous experiment has not been forgotten in the Middle East. After noting that there was some consensus of opinion on nationalisation in the developing countries, the article commented:

> This unanimity arises from the belief that the measure asserts the sovereignty of a nation over its territory and constitutes the best means of securing social justice, in that the resources thus acquired remain in the hands of the producer countries.

Although the threat of nationalisation is used as a weapon, it should be admitted frankly that motives differ greatly. The article said:

> Some have used it as a slogan, whilst others have called for it out of conviction and with deep sincerity. Despite the fact that there has not been any positive outcome, it is still a burning issue and remains

a powerful means of pressure which some manipulators – behind the scenes – do not lack skill in applying.[26]

Nationalisation is impossible while a country has insufficient facilities and skilled workers but the possibility can serve as a long-term threat or instrument of reprisal. Algeria, for example, combines reference to socialist ideology, proposing nationalisation of the means of production, with great skill in seizing the opportunities which arise. Much of the success of Sonatrach has been achieved by a series of well-timed moves. The initial aim was to obtain a broadly-based national company as quickly as possible, if necessary by usurping the assets and rights of non-nationals.

The next objective of the policy-makers was to use the state company as a base for enlarging the area captured and obtaining complete control of Algerian oil affairs. That process amounts to nationalisation.

Since a 'progressive' government came to power in Libya many moves in the direction of nationalisation of the oil companies have been made. The operators most affected are the British and American groups, because of particularly strong reaction of the military régime to the military policy of their countries, which had established important bases in Libya.[27] But the French companies have not been as violently accused of 'colonialism' as in Algeria. However, Algeria, Libya and Iraq seem to have banded together to adopt a common line of policy involving a sequence of stages ending in complete nationalisation.[28] Indubitably Algeria's action at the beginning of 1971 in taking over a majority holding in the French companies serves as a model for reference by the Arab countries.

Libyans and Algerians are not the only commentators to pay close attention to the idea of nationalism. This is put forward as a possibility by some spokesmen for the oil-producing countries.

For many years a well-known Arab petroleum consultant, Sheikh Abdullah Tariki, has been urging the producer states to nationalise.[29] He would like the oil states to reach a firm decision on this issue. If some hesitate, let the boldest take the initiative! But so far Sheikh Tariki's attitude has been applauded rather than put into practice. Many of the political leaders know that it is difficult to dispense with the foreign groups suddenly. Financial and technical dependence makes it imprudent to rush things, because of the risk of disaster. The ill-fated Iranian experiment in nationalisation has made more than one leader in a producing country think carefully and has sufficed to moderate acts of such revolutionary zeal as it has been permissible to express in official discussions. Moreover, Tarikism serves for tactical manoeuvres to come to terms with the opposing interests despite the sworn intention of kicking them out.

Saudi Arabia's Minister of Petroleum and Mineral Resources, Sheikh Ahmad Yamani, has expressed the opposite views to those held by Sheikh Tariki. In an address to the third seminar on petroleum economics held at the American University of Beirut in May 1969, Sheikh Yamani put forward interesting arguments in support of his attitude. He said that if a country decided to go it alone in nationalising in the present state of world oil affairs it was risking suicide, because of competition from other producers. He discussed the alternative that the member countries of OPEC should reach concerted agreement on taking over their national oil. What were the likely consequences? The first was that if the operation took place in a situation of over-production, then nationalisation would occur when there was a buyers' market. The second consequence would be that when prevented from making profits at the production end of the industry the majors would turn their attention to the downstream operations of refining and sale. Their interests would then inevitably be nearer to those of the consumers, in that both these parties try to buy crude as cheaply as possible. The oil-producing countries would be faced with a drop in their selling prices that they would be powerless to prevent. Should they then form a cartel acting against the consumers? Sheikh Yamani declared that 'clearly, nationalisation would be a disaster for all concerned in oil affairs – the producer countries, the consumers and the companies.' [30] These ideas help to explain why Sheikh Yamani is such a keen protagonist of the policy of participation. In 1972 he was instrumental in obtaining holdings for the Persian Gulf states in the capital of the companies operating in the Middle East. Only Iraq and Iran have adopted more radical courses.

These two contrasting doctrines help to put the matter into perspective. The first one, that of immediate and complete nationalisation, comes from a former Saudi Arabian Oil Minister, therefore from someone no longer subject to the responsibilities of power. Sheikh Yamani was, and still is, a politician holding ministerial office. The standpoints taken, by these two fellow-countrymen on the same situation illustrate Max Weber's comparison of 'the ethics of conviction' and 'the ethics of responsibility'. [31] Moreover, in order to prevail, the Tariki doctrine would need 'the means of putting policy into effect'. [32] In other words, in order to succeed in a nationalisation operation there must be a rather unlikely combination of circumstances at home and abroad. First of all a country needs to be able to assemble the human and technical resources hitherto provided by foreign companies. Then it must be in a position to find the capital required to maintain, replace and add to the facilities and equipment from the previous era. A surprise move is essential, because if no compensation is expected the departing concerns will apply a scorched earth policy. If the enterprise taking over is unable to re-start the system it will be necessary to appeal to another company for assistance. Even if the terms of the contract with this new protector are more favourable there will be a fresh set of difficulties when it comes to selling the newly-acquired oil supplies. If treatment of the companies has been too brutal there is a risk of a purchasing boycott by importers. If the oil has to pass through the network of the big companies the producer will be at the mercy of their demands. And Sheikh Yamani rightly points out that the strategy of the international groups when subjected to too much annoyance may be to concentrate on downstream operations, bringing their interests into line with those of the consumers. The effect may be to multiply opposition instead of subjugating it.

There are three essential conditions for the success of such a fundamental change: a time when demand greatly exceeds supply; a sellers' market such as there has been since 1971; simultaneous nationalisation throughout all the producer countries, or at least by all those bringing significant quantities on to the market, in order to avoid the perils of isolation.

The separate deeds must take place within a concerted strategy of all concerned, otherwise they are unlikely to have the desired results. If North Africa chooses to act separately this will not defeat the international groups, because they have worldwide scope. They could, for example, step up their activities in Nigeria, in Saudi Arabia, in the Persian Gulf or even in Indonesia.[33]

The Arab governments are well aware of all these considerations and realise that if they are too hasty they may turn forces benefiting the producer countries into agencies draining profits from those territories. Moreover, in the present climate any sign of compromise on the part of the so-called 'revolutionary' régimes may be seized upon by opponents who are only waiting for a mistake in order to take the opportunity to launch a counter-attack.

In the final analysis this is a political matter and economic rationality is not necessarily applied where political considerations are predominant. Nationalisation might be regarded as a means of extracting a country from a state of under-development but would it suffice to cure a disease that is not merely economic?

Objective of national development based on oil

The producing countries place great hopes in oil. They would like it to do more than provide revenue and satisfy domestic demand. They think it

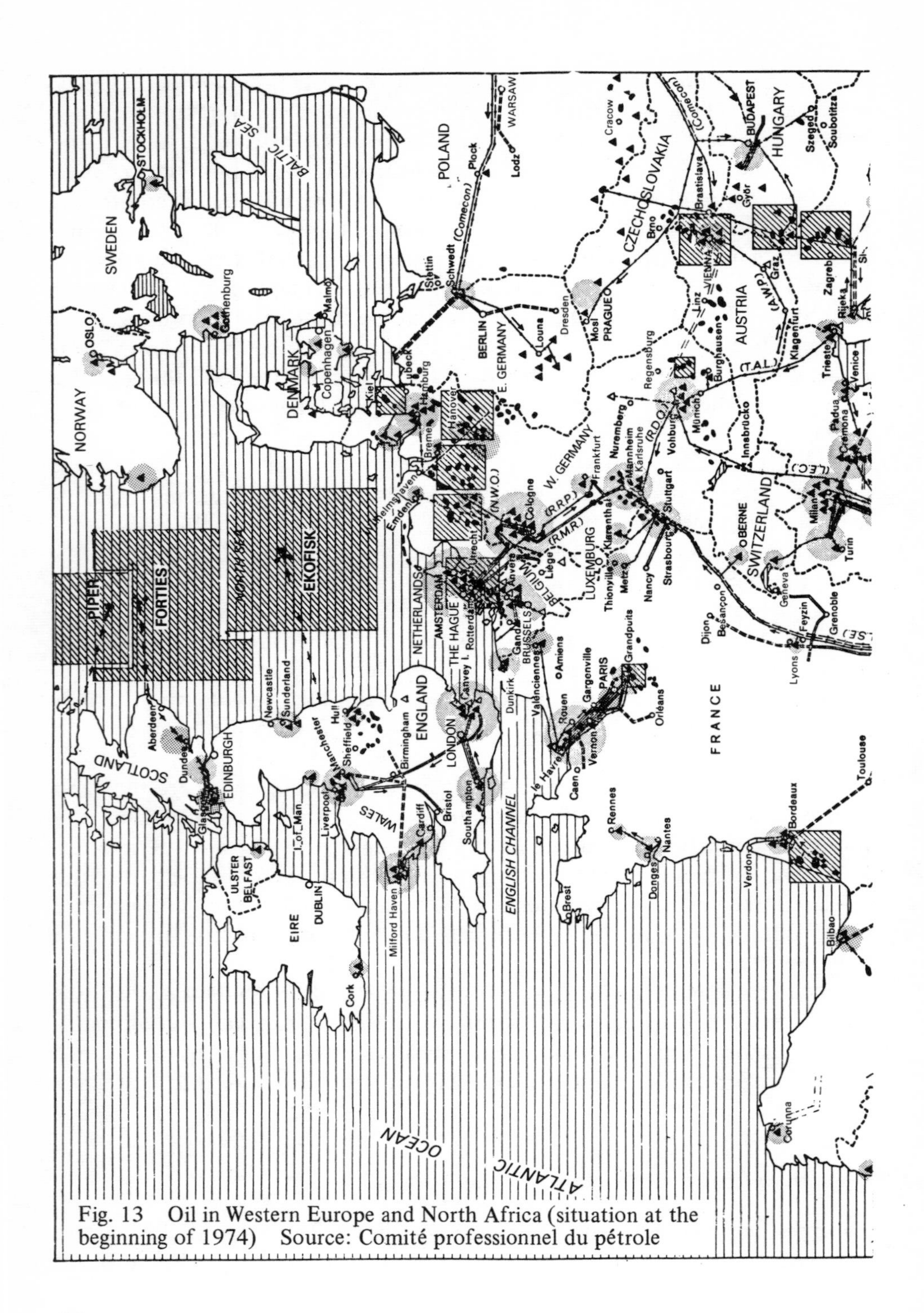

Fig. 13 Oil in Western Europe and North Africa (situation at the beginning of 1974) Source: Comité professionnel du pétrole

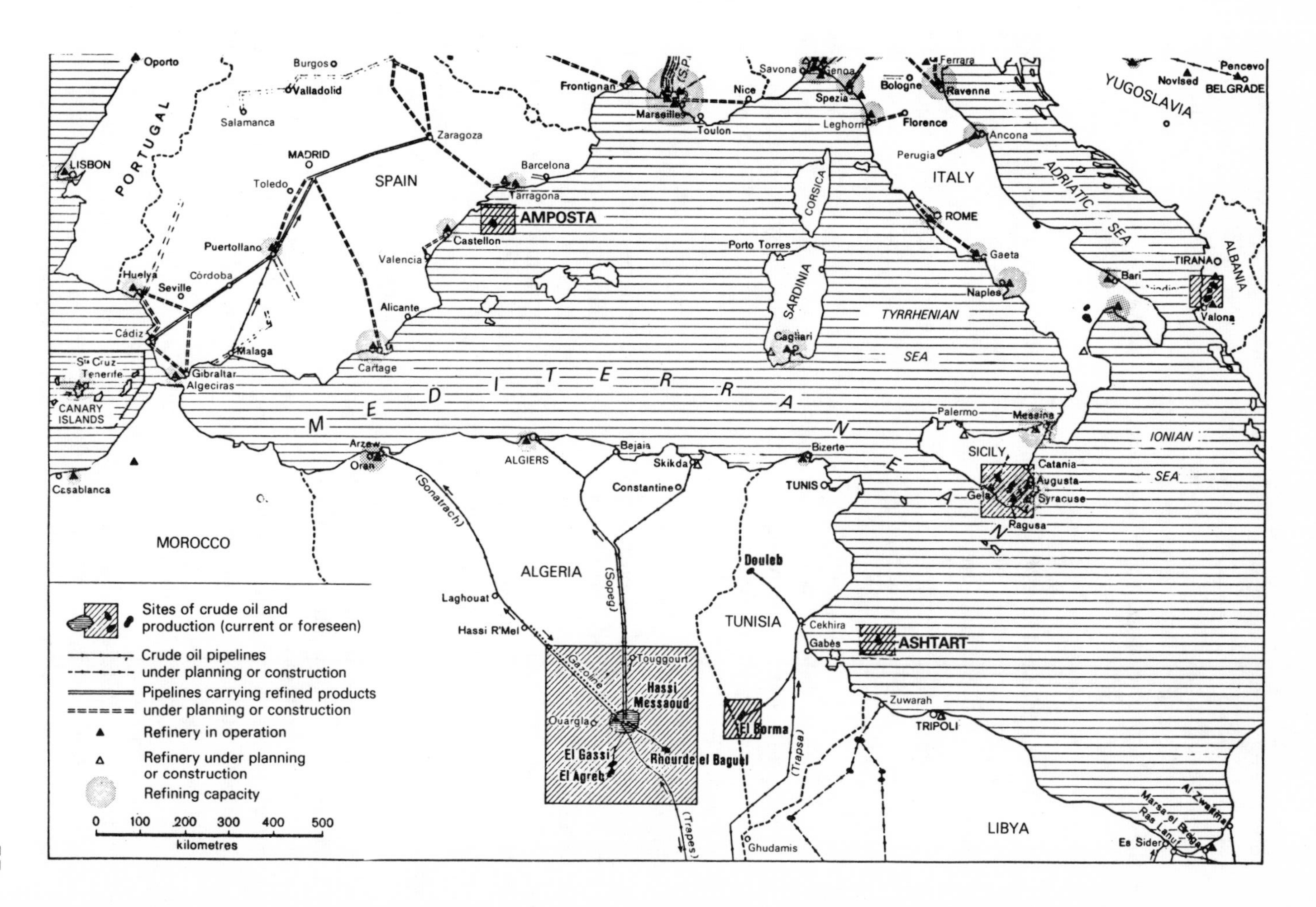

Sites of crude oil and production (current or foreseen)
Crude oil pipelines
under planning or construction
Pipelines carrying refined products
under planning or construction
Refinery in operation
Refinery under planning or construction
Refining capacity
0 100 200 300 400 500
kilometres
PORTUGAL
SPAIN
MOROCCO
ALGERIA
TUNISIA
LIBYA
ITALY
YUGOSLAVIA
ALBANIA
CORSICA
SARDINIA
SICILY
CANARY ISLANDS
MEDITERRANEAN
ADRIATIC SEA
TYRRHENIAN SEA
IONIAN SEA
Oporto
LISBON
Burgos
Valladolid
Salamanca
MADRID
Toledo
Puertollano
Córdoba
Seville
Huelva
Cádiz
Malaga
Gibraltar
Algeciras
Sta Cruz Tenerife
Casablanca
Zaragoza
Barcelona
Tarragona
AMPOSTA
Castellon
Valencia
Alicante
Cartage
Frontignan
Marseilles
Toulon
Nice
Savona
Genoa
Spezia
Leghorn
Florence
Bologne
Ferrara
Ravenne
Ancona
Perugia
ROME
Gaeta
Naples
Porto Torres
Cagliari
Palermo
Messina
Catania
Augusta
Syracuse
Gela
Ragusa
Bari
TIRANA
Valona
Novised
Pencevo
BELGRADE
Arzew
Oran
ALGIERS
Bejaia
Skikda
Constantine
Bizerte
TUNIS
(Sonatrach)
(Sopeg)
Laghouat
Hassi R'Mel
Touggourt
Hassi Messaoud
Gazoline
Ouargla
El Gassi
El Agreb
Rhourde el Baguel
(Trapes)
Douleb
El Borma
(Trapsa)
Cekhira
Gabès
ASHTART
Zuwarah
TRIPOLI
Ghudamis
Es Sider
Ras Lanuf
Marsa el Brega
Al Zwatina

should play a decisive part in the overall development of the country, causing chain reactions in other sectors. They believe that hydrocarbons could be used as a lever to get national progress moving fairly quickly, because the cash flow produced by oil can give valuable assistance towards the growth of the economy as a whole.

Views of the producer countries

In this connection the statement made by Mr Moussa Kebaili, leader of the Algerian delegation to the Seventh Arab Petroleum Congress, is illustrative of the aspirations based on hydrocarbons:

> We consider that the time has come to leave the beaten track and study clearly and objectively not only the fiscal aspect of the petroleum industry but also, above all, the part which petroleum can play in the development of our countries. This means re-examining the anachronistic system of concessions and taking charge of the exploitation of our resources. That is the only way in which we can freely dispose of our petroleum riches and use them to attain our national ambitions.[34]

(a) *Oil and industrialisation*

A Resolution of the Seventh Arab Petroleum Congress recommended the integration of the oil industry into the local economy, in order to speed up economic development.[35] In a report entitled 'The contribution of the petroleum and natural gas industry to the economic development of Algeria' Mahmoud Hamra Krouba, of Sonatrach, described the downstream operations being performed in his country: construction of oil pipelines, creation of a petrochemicals industry and building of power stations, of factories making pipes and of an iron and steel works. The underlying idea was that it was necessary to perform as much as possible of the handling of oil within the country, in order to stimulate the primary and secondary sectors of the economy and thus appreciably increase foreign currency receipts. Through oil an economy based on the export of crude products could become more diversified; the entry of manufactured goods should be balanced by the outflow of goods with an element of added value.

One of the difficulties of the operation is clearly the fact that petroleum products create industries which require heavy investment and relatively small labour forces.[36] But compared with the industrialised nations the countries which supply crude have small financial and technical resources and ample under-employed labour. Industrial employment requires semi-skilled or skilled labour. In the world market for goods there is strong competition and one of the greatest merits of that competition is that it tends to have the effect of cutting prices which are too artificial or too deliberately schemed. Therefore it will take the producer countries quite a long time to overcome their initial handicaps. They demand that in the early stages they should at least not be systematically excluded from the markets by prohibitive tariffs. They are seeking conditions in which their goods can obtain a share of the markets and receive some protection until production is on a large enough scale to permit the reduction of costs.

The producer countries want to process their products locally and sell those products in the markets of the industrialised countries. But the latter often impede those sales by maintaining discriminatory tariffs relating to the stage of processing of the products and also by imposing quantitative restrictions. A general rule which applies to petroleum products no less than to other products is that the industrialised countries prefer to use their own facilities for processing imported goods.

(b) *Oil and developing countries*

Basically the problem arises from the present limited political influence of the third world. With few exceptions, oil has so far usually been obtained in the deprived regions of the world. But because it is nowadays in great demand in countries with a high standard of living it has become one of the few products supplied by the under-developed countries enjoying a rising trend in prices or benefiting from stabilising factors. It therefore has the merit of providing one of the few opportunities for bringing foreign currency into the national treasuries. In that respect it is at present more advantageous to raise oil than to mine copper or tin, not to mention the incredible fluctuations in the prices of cotton, jute, coffee or cocoa. Moreover, petroleum products are accounting for a growing proportion of the exports of the developing countries: from 1963 to 1968 the share of hydrocarbons in their total exports rose from 30 per cent to 33 per cent, whilst that of foods dropped from 34 per cent to 28 per cent and that of raw materials from 20 per cent to 16 per cent. But one finds that this important increase only benefits a few countries, whose population only amounts to a small proportion of that of all the developing countries.[37]

Libya is included in that fortunate minority. It regularly receives the manna of petroleum royalties. Its officially declared intention is to use these funds for development. But how can the whole of the economy be assisted from this one large and growing source? The available labour force is overwhelmingly unskilled and the consumer market is very small, comprising 2 million inhabitants with an extremely low average income per head of population. Libya's development could be described as proceeding in the opposite direction from normal development. It occurs without the contribution of labour. Libya devotes exceptional attention to promoting the third sector of the economy (transport, services and building) without producing definite effects in the first and second sectors. An economist, Robert Mabro, has suggested that Libya's development might be called 'the development of a state receiving income without providing labour input'.[38] According to that writer, an oil-producing state of that nature has two essential characteristics – firstly a contrast between its poorly-equipped condition for producing goods other than oil and secondly distribution of income which neglects to give an appropriate share to means of production.

In the circumstances, there cannot be any significant results of Libya's development until the application of services and investment in the improvement of social conditions produce a lasting improvement in the value of the human resources. When that major obstacle is removed, the state receiving income without providing labour will be able to make oil the starter for national advance.

Since Libya entered the orbit of Arab nationalism this progress seems to have been its goal. But despite the size of its funds Libya feels rather ill-equipped to undertake a tough line of action under an exacting national plan. It now seems to be more inclined to look to Algeria for a rational model for development based on hydrocarbons.

The Algerian model

When Algeria began its life as a sovereign state producing oil it already had an infrastructure inherited from French trusteeship and quite a large population. It has pursued the aim of development more boldly and more comprehensively than other Arab oil-producing countries. Its demographic features and the level of national development already attained, as well as the socialist nature of its ideology have caused it to repudiate the easy-going liberal trends which have accompanied the progress of many states of the Middle East. The Algerians have noticed that hydrocarbons do not automatically yield development. Many countries producing them have become more dependent on foreigners than before the exploitation of those resources. Moreover, the standard of living has improved only for a small, privileged sector of the population. They consider that hydrocarbons should therefore be subjected to 'an effective plan of voluntarism concerning their use'.[39] That plan would endeavour to correct two basic features of all sectors of the Algerian economy – 'extroversion', i.e. the tendency to export basic raw materials unprocessed, as under the usual practice in a colonial system, and 'deep discrepancies in internal organisation', i.e. the co-existence within the country of both traditional sectors and modern sectors.[40]

The Algerian leaders take the view that 'hydrocarbons could be the specific means of creating a material basis for overcoming those circumstances, both in industry and in agriculture'.[41]

In industry they would permit the achievement of a growth of trade between the different sectors of the economy.

By the processing of hydrocarbons and the manufacture of by-products, for example fertilisers, the modernisation of agriculture could be set in train. Various equipment and supplies would be provided in order to speed up its capacity for increased production. But that promotion of expansion could be brought into effect only if crude oil and its by-products gained a place in the international market. Algeria manifests a concern for security of the market for its processed products as great as the concern of the consumer countries for the security of their supplies. It regards exclusion from the markets as being as damaging to its interests as a stoppage of supplies is to the importers.[42]

Putting those ideas into practice, Algeria has undertaken a development programme based on the oil industry. It seeks to expand the foundations of the national economy by setting up basic industries promoting rationalisation of production and profitability, by reorganising the structure of agricultural production with a view to commercial success and by rapid training of the personnel required to provide the nucleus for progress in industry and agriculture.

The planning calls for simultaneous use of profits from hydrocarbons to finance the other sectors and investment in hydrocarbons in order to achieve rapid increase in their profitability. Thus under the present plans development of the economy centres on a kind of gamble based on hydrocarbons.

Whilst it is possible to envisage that the economy can be given impetus by this means, it should be noted that apart from the question of how the funds for financing the various projects are to be obtained there are two possible threats to that process. Firstly there is the difficulty of placing Algerian petroleum products in an international market which is already highly organised and expanding vigorously. For example, refining operations tend to be transferred from the production areas to the consumer areas, leaving the producer countries with so-called midstream refining only.[43] Moreover, in present circumstances, the scope of the effects created by oil operations is limited. Thus, it is a definite fact that so far oil production has had slight influence on the expansion of agriculture. Yet 70 per cent of Algerians still depend on agriculture.

The efforts of the oil-producing states to achieve development are encountering serious difficulties, despite the anxiety of their leaders to obtain rapid progress. The main obstacles prove to be shortage of skilled labour and the inadequate technical facilities. Finance is not a key factor, as it depends on the confidence created abroad and the size of the funds spent on promoting exports. This assistance, in turn, depends on the volume of oil extracted and on the prices which potential buyers are prepared to pay. Although they are in a better position than other nations of the Third World to finance investments, the oil-producing states are beset by more serious ills arising from inertia and sociological factors, preventing a go-ahead attitude from being created quickly. That is also the reason why they fail to strike out on independent paths of action but are affected by foreign domination from the wealthy countries, including socialist states. The possession of raw material is not a guarantee of success. There is no magic formula, however much journalists repeat that term. In fact, the real power lies with the consumer countries at present, because of their immense purchasing power, their capacity for organising to protect interests regarding which they are increasingly tending to adopt collective attitudes, their advanced level of scientific achievement and the amount of technical expertise at their disposal. Countries which import crude oil in large quantities more than make up the expenditure on these massive purchases from the added value obtained with the processed products. The initial raw material is only a starting point for a set of end products worth a great deal more because of the skilled labour and capital incorporated in them. Through its oil operations a producing country is drawn into a position deeply affected by the insatiable nature of the demand from better-placed importers.

Thus, the producer countries, and specifically Algeria and Libya, pursue several aspirations simultaneously or in turn. This chapter has discussed the three which seem to be the most characteristic – those of increasing oil revenue, of 'naturalisation' of the country's oil and of using oil as a basis for economic development.

It is now appropriate to consider more carefully whether or not the objectives of the producer countries can be reconciled with those of the consumer countries. Are the two sides enemies with uncompromisingly antagonistic attitudes or are they potential allies with differing but relatively complementary interests? Is the relationship between these two participants in oil affairs one of unalterable opposition, of peaceful co-existence or of willingness to come to terms despite occasional quarrels?

Notes

[1] OPEC (Organisation of Petroleum Exporting Countries). Launched in September 1960, in Baghdad, OPEC is an intergovernmental organisation which has been approved by the United Nations. In 1974 there were twelve members: Abu Dhabi (since 1967), Algeria (since 1969), Saudi Arabia, Ecuador, Indonesia, Iraq, Iran, Kuwait, Libya, Nigeria (since 1971), Qatar and Venezuela. Gabon has been an associate member, without voting rights, since November 1973. The inclusion of Trinidad is expected. The admission of new members requires unanimous approval. A condition of membership of this consultative body is that 'substantial' oil exports are made. The organisation, whose headquarters were transferred in 1965 from Geneva to Vienna, has a ministerial conference with one delegate from each country, an executive council and a general secretary who holds office for three years with the assistance of a secretariat comprised of five divisions – administrative, economic, legal, technical and information. There are also two special units, an economic commission set up in 1964 and a co-ordinating committee for national oil companies which was formed in 1968.

[2] OAPEC (Organisation of Arab Petroleum Exporting Countries). This was inaugurated in 1968 by three leading Arab oil-producing countries, Saudi Arabia, Libya and Kuwait, and was joined by Algeria, Bahrain, Abu Dhabi, Dubai and Qatar. Iraq, Syria and Egypt were admitted at the beginning of 1972, after the articles of association had been revised to provide that oil need not be the 'principal and basic' source of revenue but could be merely an 'important' source of income from exports.

[3] In 1970 the members of OPEC held 64 per cent of world reserves and 48 per cent of total production was in their territories; their exports comprised 85 per cent of world exports.

[4] P. Clair, *L'indépendance pétrolière de la France. I, Le théâtre de guerre,* op.cit., p. 107.

[5] P. Clair, 'La structure des prix postés pétroliers. Perspectives et statistiques.', *Revue économique,* May 1966, pp. 467 et seq.

[6] *Petroleum Press Service,* May 1970, p. 186.

[7] A. Murcier, 'Un expédient provisoire? Le pétrole entre l'Algérie et la France', *Le Monde,* 21 March 1969.

[8] *Marchés tropicaux,* 1256, 29 November 1969, p. 3107.

[9] According to *Petroleum Press Service,* Algeria's oil revenue in 1969 was 255 million dollars and Libya's was 1,132 million dollars from three times the volume of production in Algeria. See *Petroleum Press Service,* September 1970, p.324.

[10] 1969 revenues were: Libya 100 cents per barrel; Saudi Arabia 87·1; Kuwait 80·8; Iran 81·5; Iraq 91·4. Overall mean for 1969 for all the countries of the Middle East: 84·7. From all the countries of the Middle East plus Libya the mean obtained is 87·8 cents per barrel in 1969. Only Venezuela, with a revenue of 103·6 cents per barrel, is above Libya in the table. See *Petroleum Press Service,* September 1970, p. 325.

[11] P. Clair, op.cit.

[12] P. Balta, 'La France et le monde arabe. I. Les réalités économiques', *Revue de défense nationale,* May 1970, p. 777.

[13] M. Byé, 'La grande unité interterritoriale', op.cit.; cf. P. Clair, op.cit., p. 110.

[14] Ahmad Zaki Yamani, 'Participation versus nationalisation, a better means to survive', *Middle East Economic Survey,* XII (33), 13 June 1969, supplement.

[15] Libyan production in 1970 was slightly higher than in 1969. The Government forced the companies to reduce their output while they failed to accept the new tax demands made since the change of régime. The figures for 1972 and 1973 show a decline from the 1969 level.

[16] J. Masseron, *L'économie des hydrocarbures,* op.cit., p. 82.

[17] Ibid., p. 82. At the time of the opening in Algiers of the Twentieth OPEC Conference, ERAP and CFP were asked by a letter from Mr Abdesselam, Algerian Minister of Industry and Energy, to repatriate to Algeria 'an amount at least equal to the equivalent of 1·80 dollars per barrel of crude oil exported'. On the basis of the price of oil at that time, this demand meant making the two French companies remit to Algeria 95 per cent of the proceeds of their exports. See also *Marchés tropicaux,* 1286, 4 July 1970, p. 2114.

[18] OPEC Conference, Vienna, June 1968, Resolution XVI–90, Cl. 2.

[19] Ibid., Resolution XVI–90, Cl. 1.

[20] *Middle East Economic Survey,* XIII (26), 24 April 1970.

[21] Ibid.

[22] Secretariat Social d'Alger, 'Le pétrole en Algérie', *Bulletin mensuel,* September 1967, p. 22.

[23] Sixth Arab Petroleum Congress, 13 March 1967: 'The Congress recommends the strengthening of national oil companies and organisations operating in Arab Member States and calls for the establishment of similar

companies and organisations in those Arab countries where none exist at present.' Cf. *Selected documents of the international petroleum industry 1967*, edited by Nameer Ali Jawdat, OPEC, Vienna 1968, pp. 273–4.

[24] KNPC and Hispanoil have holdings of 51 per cent and 49 per cent respectively in Kuwait Spanish Petroleum Company, which has a 9,000 square kilometre concession in Kuwait.

[25] 'Kuwait Congress', *Petroleum Press Service,* May 1970, pp. 177–9.

[26] 'Les pays arabes sont-ils en mesure de nationaliser l'industrie pétrolière?', *Bulletin de l'industrie pétrolière,* 16 April 1970.

[27] See Chapter 7.

[28] According to the communiqué issued on 23 May 1970, following the meeting in Algiers of the Ministers responsible for Algerian, Libyan and Iraqi oil affairs, the three countries '. . . emphasised that the rational use of oil resources represents the best weapon for the attainment of development and economic independence. To this end they declared their determination to strive for the integration of oil operations within the national economy in order to endow with all the prerequisites of success the efforts being undertaken by the three revolutionary powers to fulfil the legitimate aspirations of their people for a better life. The three ministers consider that the best guarantee of success lies in the existence of an authentic national industry in all the principal phases of oil operations, representing the first stage of direct exploitation and full governmental control over the national resources . . .'. *Middle East Economic Survey,* XIII (31), 29 May 1970, p. 2. A few weeks later, in an interview given on 5 July 1970 to the *Financial Times* correspondent in Tripoli, Mr Mabruk, Libyan Minister of Petroleum and Minerals, declared, '. . . So long as the relationship between the companies and the government remains harmonious and the companies realise the changing circumstances and continue to co-operate effectively with goodwill in solving our outstanding problems, I don't think we need nationalisation . . .'. *Middle East Economic Survey,* XIII (37), 10 July 1970, p. 6. It has been said that Mr Mabruk was personally opposed to Coloret al-Qadhafi's proposal that the British company BP should be nationalised.

[29] A. Tariki, 'Nationalisation of Arab petroleum industry is a national necessity', Fifth Arab Petroleum Congress, Cairo, March 1964.

[30] A. Yamani, 'Participation versus nationalisation, a better means to survive', *Middle East Economic Survey,* XII (33), 13 June 1969, supplement.

[31] M. Weber, *Le savant et la politique,* Plon, Paris 1963, pp. 172 et seq. (Collection 10/18).

[32] P. Clair, op.cit., pp. 110–11.

[33] Indonesia has a very promising oilfield and political conditions there are now favourable to the foreign companies.

[34] *Bulletin de l'industrie pétrolière,* 1552, 25 March 1970.

[35] On recommendations made by the Seventh Arab Petroleum Conference, see *Petroleum Press Service,* May 1970, pp. 177–9.

[36] H. Dangeard, *Localisation européenne des grandes unités pétrochimiques,* Ed. Technip, Paris 1970, pp. 79–107 (Colloques et séminaires No. 15).

[37] GATT treats as 'major oil-exporting countries' the producer countries of the Middle East, Libya, Venezuela, the Netherlands Antilles, Trinidad and Tobago, and Brunei. But according to the GATT report published in 1967, 'the total population of that country comprises less than 3 per cent of the overall population of the developing countries. And the exclusion of Iran, which in oil exports accounts for only 15 per cent of the total, although its population constitutes 45 per cent of that overall population, gives an even higher concentration: less than 1·5 per cent of the population of the developing countries provides about 25 per cent of the total exports'. See *International Trade in 1966,* GATT, Report, Geneva, p. 248, note 1. The figures for change in percentages by class of goods are given in *International Trade in 1969,* 1970, pp. 22–3.

[38] R. Mabro, 'La Libye, développement d'un Etat rentier?', *Projet,* 39, November 1969, pp. 1090–101.

[39] S.A. Ghozali and G. Destannes de Bernis, 'Les hydrocarbures et l'industrialisation de l'Algérie', *Revue algérienne des sciences juridiques, économiques et politiques,* VI (1), March 1969, p. 272.

[40] Ibid., p. 257.

[41] Ibid., pp. 292–3.

[42] Ibid.

[43] P.H. Frankel and W.L. Newton, 'Current economic trends in location and size of refineries in Europe,' 1959. This study was the basis for a discussion arranged by the Institute of Petroleum of London.

3 Opposing Interests

The consumer countries seek to obtain assured supplies of oil at a reasonable cost; the producer countries strive to supply oil at a profit, with the proceeds coming back to the nation and bringing about its development. The interests of these two sides differ from the outset, because the former approach matters from the direction of demand and the latter from the direction of supply. But despite this contrast they meet in a commercial transaction, precisely because the buyers are looking for supply and the sellers are seeking demand. Regulation of supply and demand is effected by fluctuation of prices. According to correct economic theory the intersection between the supply curve and the demand curve gives a price at which trade between counterparts can take place. But this concept supplied by economics is too highly schematised: an objection is immediately apparent, namely that oil prices are not dictated by economics but are greatly influenced by politics, above all in the countries with Mediterranean coastlines. The answer to that complication is that in order to judge oil matters adequately it is necessary to take into consideration the whole agglomeration of political influences and economic factors. The fact that political motivations affect the interplay of various interests in the oil market does not make the economic law of supply and demand inapplicable: it merely means that the law is deflected according to the power of each of the participants. In order to have a chance of approaching its goals each contender must give a credible impression of having decided the precise outcome which it is seeking at national or international level. Although the decision taken expresses the power relationship existing at any given moment, it can always be translated into economic terms in the form of a relatively measurable cost. Political constraint, which operates as if it determined a quantitative level on a graph, adjusts the point at which there is some balance between the aspirations of the participants.

In the countries with Mediterranean coastlines, to which oil trade is such an important matter, politics and economics mingle where hydrocarbons are concerned. The objectives of the camps involved are by and large antagonistic: one side wants cheap oil and assured supplies, the other side is trying to obtain a high price for oil and to impose national control of it. The solution might be to remove one of these antagonists from participation. However, events are not moving in that direction.

In this book an attempt has been made to explain the state of affairs. It has been postulated that neither side can obtain a decisive victory over the other by unilateral means. Certainly the divergencies are great, but there is also considerable scope for closer alignment. In studying the encounter between the objectives of the countries of this geographical area it is appropriate to consider the effects of both irenicism and permanent hostility, in order to assess the combination of affinity and discord.

Discussing relations between the United States and the USSR, Raymond Aron took up the theme of 'warring brothers', which can be transposed to the context under review here. Mr Aron commented in his book *Peace and war between nations* that the two great powers were adversaries because of a 'hostility of position',[1] but that economic progress engendered a 'growing fraternity'.[2] Suppliers and purchasers are separated and drawn together in a similar manner. Mr Aron's observations regarding countries which are 'warring brothers' may be applied to relations between buyers and sellers of oil in the Mediterranean area. These participants are in antagonistic positions in the scheme of oil operations but the development of trading relations imposes rapprochement:

> The idea that the two Great Powers of an international system are brothers as well as enemies should be regarded as normal rather than paradoxical. By definition, each would reign alone if the other did not exist. Now, candidates for the same throne always have something in common. The units of an international system share a common area of civilisation. Inevitably, they espouse the same principles to some extent and conduct a debate whilst waging a combat.[3]

The effect of these paradoxes in the case of the subject under study will now be discussed. In the oil affairs of the countries either side of the Western Mediterranean a duel occurs but it cannot be fatal to either of the participants. The participants are, in fact, 'rival associates', whose strategies are complex; they are participants in a conflict situation, at the same time engaging in a kind of co-operation between contenders.

'Rival associates'

Participants on either side of the Mediterranean spend the time challenging each other and even trying to eliminate each other, but it is impossible to achieve decisive elimination. The bitterness of the competition should not prevent observers from bearing in mind the extent and depth of elements of accord. Hence the relations between the producers and the consumers may be viewed either from the point of view of rivalry or in terms of co-operation. Or, in political terminology, it is possible to emphasise either the amount of dissension or the degree of consensus. In order to achieve precision it is appropriate to turn to an abstract scheme which enables this paradoxical situation to be evaluated and which is applicable in both the political and the economic spheres. It should be one that fits the collective behaviour of several participants who are capable of separate decision-making (polyarchy).

A comprehensive model of the kind sought seems to be contained in the commentaries of François Bourricaud, whose thinking has been influenced by American writings on the topic, notably those of Robert A. Dahl, to whose ideas he refers frequently. In his book *Esquisse d'une théorie générale de l'autorité,* Bourricaud attempts to see how power relations function in a polyarchical system. To characterise the continuous interplay of interest groups endeavouring to influence each other, to oppose each other or to form coalitions, the writer refers to 'symbolic manoeuvering' among 'rival associates'.[4] The participants are engaged in measuring each other up, performing trials of strength, or even sometimes trying to eliminate other players – but without ever achieving that result. The intensity of the conflict dividing them is counterbalanced by the strength of the bonds uniting them.

Complex strategies

Thinking on economic policies in oil matters has long been primarily concerned with the omnipotence of the large international groups known as the seven majors.

Since the large international companies are of fundamental importance in oil affairs, this terrain might be expected to be a fruitful source of new ideas about oligopolies.[5] The situation of competition in a context of association links up with the reasoning of a number of economists who have studied situations of imperfect competition, in which a small number of firms dominate the market. Many writers, among them E. Chamberlain, have emphasised the ambivalence characteristic of the sales policy adopted by oligopolists: there is independence, because each seller seeks maximisation of its own gain, but there is also 'mutual dependence', or interdependence, because of the occurrence of explicit understandings, or tacit avoidance of provoking hostility, having regard to the losses which prolonged rivalry would cause. In fact, the dominant contender has to reflect on the indirect consequences of its policy of maximisation of profits.[6]

But neo-classical theories concerning oligopolies have three important shortcomings where the subject-matter of this book is concerned.

Firstly, they are chiefly concerned with the sales tactics and price policies of the oligopolists. They treat demand as being above all a matter which the decision-makers have to take into account in their thinking. But in oil policy a more complicated demand situation has to be dealt with. The buyers are powerful groups who do not act individually but collectively, so that a small number of buyers appear in the market, presenting a counter-strategy to that of the sellers. Oligopolistic theory then has to be so designed that it can relate to oligopoly in both supply and demand.[7]

Secondly, the State plays an important part in connection with new policies in oil matters, by putting into effect general regulations and limiting legislation applicable to the activities of private enterprise. It also creates public corporations whose behaviour is influenced by economic considerations but which must also take into account in their decisions the overall political intentions of their sole proprietor.

Thirdly, those theories pay too much attention to the structures of an oligopolistic market and not enough to behaviour in that market. Yet a firm must constantly anticipate possible reactions.

Internally an enterprise is subject to certain limitations as regards strategy, its overall policy being governed by a degree of prudence.[8] In relation to the external environment an enterprise endeavours to adapt itself to the behaviour which it expects its competitors to adopt. It is constantly considering how its rivals are likely to react if it makes one specific decision or another regarding prices or quantities or investments. It makes that choice of action which is best designed in the light of the way in which matters may develop in the course of time.

Games theory seems to offer the closest approximation to the real situations in the field under consideration. Its ideas dealing especially with semi-cooperative games provide useful guidance in that respect. It is concerned with expressing interconnections between an economic goal, the desire for gain and human actions having aims wider than material benefit alone. Strange as it may seem, games theory is above all able to provide enlightenment regarding the areas of uncertainty and kinds of relationships which this chapter is seeking to investigate. It permits explanation of why collaboration occurs in a context of rivalry, since it reveals that the number of possible actions and the range of choices within each possibility are not infinite.

As an exponent of this theory, A. Rapoport explains:

> However many strategies there are, the number cannot be infinite so long as the number of possible moves and the number of possibilities at each move remain finite. Moreover, it should be appreciated that when each of the players has chosen a strategy from among the possible strategies, then the course of the game is entirely pre-determined. This does not mean that each of the players knows from the start how the game will go. He cannot know this if he does not know his opponent's strategy as well as his own. But an observer who knows what pair of strategies has been chosen (one for each of the players), can in principle deduce from this the whole course of the game.[9]

Thus, for example, Algeria and Libya are in competition with each other as oil producers; they are also potentially under competition from other oil producers. They may decide on individual strategy or they may seek coalition: they will decide on coalition if they expect it to yield a greater gain than the sum of the gains which each of them might expect from acting separately. But the final result will depend on the conflict situation. If they are engaged in a 'zero sum game', then no coalition can benefit both players simultaneously, but if a 'non-zero sum game' is involved, then the outcome may benefit each of the two players who decided at the start of the game to act in collaboration.

Games theory can provide a useful model for explaining the competition between sellers or buyers of oil who are on the same side, by representing each as a player who has a specific position of his own but who is involved in one and the same socio-economic game. Yet I do not think it can give a complete explanation of all the economic and political relationships arising in oil affairs. Complications arise from the fact that in oil matters there are three main participants and they are obliged to adopt strategies conditioned by the three-sided nature of the contest. For example, the suppliers have to make their decisions according to circumstances in connection with potential buyers and also in the light of policies of the governments of the producer and the consumer countries. Furthermore, the three-sided relationship is only a frame of reference and beyond it there are additional sets of conflict situations: the actions of the sellers and of the buyers may in turn be affected by other actual or potential sellers or buyers.

Conflict situation

It has already been seen that there are three main contenders involved in the interplay of economic and political relationships around oil. They are the consumer countries, the producer countries and the companies. Whether the companies are private or public enterprises they function as intermediaries.

The interests of the three participating sides are not identical, hence the rivalry occurring in contacts between them. In the consumer countries the state companies pursue policies differing from those of the private companies, and nowadays governments prefer the former to the latter. In the producer countries the governments are confronted by international capitalist groups and resolutely adopt a controlling attitude, sometimes by allying themselves with foreign companies operating within the nation, sometimes by using for support national companies created in their country. The 1965 oil agreements between France and Algeria sought to deal with the matter directly at a political level, by obliging the companies to operate within a system prescribed by an actual treaty under international law. But although clashes were momentarily submerged by apparent understanding, they soon recurred as vigorously as ever. The Algerian Government then kept calling on the French Government to force the French companies to accept the conditions imposed by the Algerians. But the French Government, for its part, could not be indifferent to the plight of its companies, even if its horizons were wider. Libya has used Noc and agreements with French or other foreign companies as a means of putting pressure on the private oil enterprises, the majors and the independents, which were allowed by a régime that is now abhorred to establish themselves in the national territory.[10] That confrontation is constantly shifting between bipolarity and multipolarity.

Is the state company capable of serving as a special means of bringing the opposing interests of purchaser and vendor governments into accord with each other? The Algerians seem to doubt its ability to do so. They regard oil operators of any nature as 'exploiters'. The newspaper *El Moudjahid,* which is very much the voice of the régime, has declared that there must be greater collaboration between progressive producer countries, in order to 'countervail the unity existing among the exploiting forces of oil operators'. That newspaper has stated:

> Whatever some people would have us believe, one thing is certain – that if certain companies such as ERAP or ENI boast of not belonging to the cartel their divergence is very slight. Competition

> between them only comes into effect in matters of distribution or of European outlets: these differences occur solely in connection with their interests in the West.
>
> But as soon as it comes to adopting an attitude towards the producer countries the traces of division disappear; the positions are similar, if not common to all those concerned. The same methods of deriving the maximum profit from oilfields are employed, to the consistent detriment of the rightful owners of these riches.[11]

It is true that the French companies assumed a manner of behaviour in Algeria that was similar to that of the international groups in Libya. Any difference was attributable to the fact that they carried less weight and their strategy did not have such an extensive worldwide scope. Acting as allies in effect, despite different approaches, all the 'Western' oil operators seek to curb the rise in posted prices and to keep profits which render their work worthwhile. The situation obliges the various companies to follow similar lines of action. But as soon as the situation becomes less tense each operator tries to return to its own choice of conduct. In that matter, as in other respects, armistices can only be temporary. Even so, conflicts between companies, whether they are private or public enterprises, are less profound than those between buyers and sellers.

Co-operation between contenders

This conflict situation should not, therefore, be treated as one involving two sides.

In addition to opposition between sellers and buyers there are other rivalries which are less evident but equally important. In the international scheme there are divisions cutting across the producer and consumer camps. Algeria is well aware that, for example, although it shares the ideology of socialist revolution with Libya, it is in opposition to Libya on various matters concerning oil. Moreover, even if the two leading oil-producing countries of North Africa co-ordinated their oil policies, thereby achieving hardening of their line in relation to the West, other producer countries might gain from this unexpected situation and make advantageous agreements with the European countries, irrespective of ideas of solidarity between producers. In any event Madrid, Rome and Paris cannot individually subscribe to purchasing arrangements at unduly high prices because this would lead to the whole of their industrial system being handicapped in international competition with the other developed countries. Accordingly demands which are too heavy must constantly be scaled down in order to gain acceptance.

Hence, an extreme proposition may be envisaged but is hardly likely to become a reality. Behind the pattern of rivalries which is the most prominent feature of the scene there is implicit collusion between participants. The fortunes of the producer countries are bound up with those of the other contenders through the proceeds of operations and the need to finance exploration and production. Once these realities are perceived, a kind of co-operation between contenders arises. Again, François Bourricaud explains the point, describing the tensions and armistices occurring in relationships between 'rival associates'. He points out that 'rival associates' are condemned to living with each other despite their divergent economic interests and although their desires in connection with political power are irreconcilable. Each contender pursues its own advantage and its specific ambitions but none is strong enough to impose its point of view completely and the sole solution open is for it to opt out of the game and be replaced by other players.

According to N-person game strategy, which serves to illustrate the point, 'the possibility of coalition does not pose new problems for the players, except perhaps regarding the choice of a system of values'.[13]

On both sides of the Mediterranean the system of values is, clearly, stretched to the maximum, but exponents of games theory do not require the final choices of the players to be perfectly co-ordinated. They consider that a coalition is viable if the system of values is guided, as appropriate in the specific circumstances, by maxims of the general interest, of stability and of effectiveness.

In fact, stability is often in jeopardy and that is the reason why strategies of encounter between the groups of countries on either side of the Mediterranean shift between strategies of co-operation and strategies of rift. However, it seems to be difficult to attain the extremes in practice, hence collaboration without conflict is just as unlikely as the other extreme of fatal menace. The extreme threat is that of death or that of suicide. As a commentator has stated, 'A threat becomes a threat of death if it ruins the adversary whatever action he chooses. It is a threat of suicide if the adversary has at least one line of action available to him which causes the ruin of an opponent threatening him in all possible circumstances'.[14] In the case of the countries with Western Mediterranean coastlines, a threat from a member on one side to a member on the other never goes to that extreme. It is at an intermediate level determined by how much gain the threatener expects to achieve at the expense of the foe in trying to escape from the previous state of affairs. Between producer countries and consumer countries on either side of the sea there is some room for manoeuvre between limits which are set short of complete agreement or all-out war.

Accordingly each particular strategy must allow for the moves made by the other side, anticipating its reactions in order to be better able to influence or countervail them, and it must eliminate from its own propositions anything that is too fixed or too unrealistic. [15] As in games theory, each participant acts according to its own goals and taking into consideration the foreseen reactions of the opponent or opponents. Collective organisations such as governments, private or public corporations, commissions or negotiating bodies constantly have recourse to indirect methods. Bourricaud asserts that,

> The art of government largely lies in skill in waiting and making others wait, in holding back in order to avoid commitment to a particular action and trying so far as possible to arrange that a 'rival associate' bears the responsibility for measures which are necessary but unpleasant, or at least engaging oneself only when protected by the support or the neutrality of all those who could subsequently find in one's present initiatives a means of creating an arrangement to one's detriment.[16]

The ferocity of rivalry depends also on inhibiting factors preventing 'rival associates' from taking the maximum advantage of a state of affairs momentarily favouring them. The degree of pressure exerted by a player depends on the type of confrontation chosen, for example whether debate is taking place *in camera* or before public opinion which is informed of the proceedings but may be imperfectly aware of the issues and the factual details. In any event, no participant can obtain a decisive victory because that would destroy an implicit consensus. In reality, in circumstances of a multiplicity of decentralised powers of action every participant needs the others, whilst seeking to obtain the best possible gain at the expense of others. According to Bourricaud:

> Two images for obtaining an impression of the functioning of polyarchies are available. Sometimes rivalry is described as a fight to the death, as a desperate struggle between two sides. That image owes a great deal to a certain interpretation of Marxism. But a second representation is available, pointing out oblique and prudent steps and going no further than the simplest and most surreptitious forms of rivalry. The other participants in the relevant affairs are 'associates', because none can achieve anything without the others;

they are 'rivals' because the gain by each is to the detriment of the others.[17]

Hegel pointed out long ago that dialectics between master and slave tend towards recognition by each side of the position of the other, because if one of them eliminated his opponent completely he could not become truly himself.[18] I believe that, because of their flexibility and their capacity for gathering together most of the variables actually occurring, these theoretical formulations enable an explanation to be given of the attainment of conciliation between the divergent objectives contained in the overall agglomeration of oil affairs in the countries on either side of the Western Mediterranean.

The producer countries and the consumer countries are in this situation of rivalry with each other which does not exclude phenomena of co-operation. The means of achieving rapprochement may be strengthening of the state companies on either side. The national companies of each camp have opportunities for meeting up with each other because they journey in opposite directions to each other, the companies of the producer countries are proceeding downstream from the well to the pump and those of the consumer countries are going upstream from the pump to the well.[19]

In order to be able to develop public corporations, France, Italy and Spain are obliged to limit the involvement of private oil enterprises in their countries. For the purpose of achieving a public oil sector, Algeria and Libya have to limit the scope of action of foreign concerns. Between the Matteistic policy of the purchaser governments and the policy of national socialism of the vendor governments an alliance is possible. If alliance between those sides is possible it might eventually imperil the equilibrium of the international private groups by dangerously restricting their possibilities for operations.

The private groups have less latitude than the public concerns for accepting the expression of matters in political terms. But rivalry in the economic sphere could potentially be offset by desire for understanding and pursuit of co-operation at a political level. Arrangements in oil affairs would then take place in the enlarged setting of chosen policies for international contact between two countries. Relationships between the various participants in oil matters would necessarily be modified. Considerations of overall policy would then extend over an oil policy deprived of its distinctive features. Oil would decisively cease to be treated as an area of activity chiefly within the orbit of economics and would be transferred to politics in a position governed in turn by international politics.

Notes

[1] R. Aron, *Paix et guerre entre les nations,* Calmann-Lévy, Paris 1962, p. 534.

[2] Op.cit., p. 539.

[3] Op.cit., p. 527, beginning of chapter XVIII, entitled 'Les frères ennemis'.

[4] F. Bourricaud, *Esquisse d'une théorie générale de l'autorité,* Plon, Paris 1961, especially chapter IV, 'Les associés-rivaux', pp. 319–52.

[5] H. Mercillon, 'Nouvelles orientations de la théorie de l'oligopole', *Revue d'économie politique,* January–February 1961. See also H. Guitton, 'Les modes nouveaux de formation des prix. Prix avec et sans marché', *Revue d'économie politique,* 3, May–June 1967, pp. 269–98, and *Economie appliquée,* special issues April to September 1952 and July to December 1955.

[6] E. Chamberlain, *The Theory of Monopolistic Competition,* Harvard University Press, 1950, pp. 46 et seq.

[7] P. Clair, op.cit., p. 46

[8] 'Les stratégies de l'entreprise', *Economies et sociétés,* II (3), March 1968, special issue.

[9] A. Rapoport, *Théorie des jeux à deux personnes,* translation by V. Renard, Dunod, Paris 1969, p. 27. (Published in English as *Two-person Game Theory: The Essential Ideas.)*

[10] In 1973 ENI was used as a tool for this policy.

[11] *El Moudjahid,* 25 May 1970.

[12] F. Bourricaud, op.cit., p. 325.

[13] C. Berge, *Théorie générale des jeux à N Personnes,* Mémorial des sciences mathématiques, Paris, Gauthier-Villars, 1957, pp. 83–4.

[14] M. Schubik, *Stratégie et structure des marchés,* Dunod, Paris 1964, p. 199.

[15] Ibid., pp. 335–6.

[16] Ibid., p. 333.

[17] Ibid., p. 329.

[18] Hegel, *La phénoménologie de l'esprit,* vol. I, Aubier, Paris 1939, pp. 160 et seq.; cf. p. 160: 'But this supreme proof by means of death eliminates precisely the truth which should emerge from it and at the same time certainty of oneself in general'.

[19] P. Clair, op.cit., p. 114.

4 Evolution of the State Oil Companies

A theme running through the previous chapters of this book has been the question of who is winning in the oil game, so far as the specific geographical area under special consideration is concerned. Which of the contenders is doing best? Is it the consumer countries or the producing states, or is neither of these sides receiving the most benefit? Are the big companies or the state companies prevailing in the struggle? Can there be any definite overall winners? Is politics capable of applying morals or methods that regulate these conflicts arising in the economic field? Is it possible to dispense with the concept of co-operation between competitors and view oil matters in terms of complete co-operation or outright competition?

An attempt will now be made to answer these fundamental questions. In order to do so it is necessary to consider whether, in the circumstances, states are trying to achieve too much or whether the means are suited to the ends.

Discussion so far has pointed towards the fact that, broadly speaking, there are three types of approach by states to oil affairs – the liberal model, the model of general regulation and the model of direct management.

The liberal model consists of *laissez-faire,* i.e. the principle of non-intervention of the Government in contests between private interests. The petroleum deposits go to whoever makes the most advantageous bid and anti-trust legislation corrects the worst excesses of capitalist concentration. Whilst liberalism is often proclaimed in public statements, the fact is that it is rarely practised where oil is concerned. Some people would like to see it applied. One should consider the advisability of non-interference by public authority in private economic activities.

The model of general regulation is the one most often encountered, because in the oil business interventionism is the rule and liberalism the exception. Indirect intervention by the State occurs in decisions by its parliament, its government, the administration and various committees or officially recognised bodies, where those decisions reflect a policy, guiding efforts in a particular direction and with a definite purpose. Such decisions, contained in laws, decrees, orders and instructions, lay down a system within which individuals, groups or bodies may conduct their affairs. They are indirect in the sense that the intention is to oblige those concerned to do or to refrain from doing what the authorities have judged to be beneficial or harmful to the general interest. They may be regarded as a means of urging, encouraging, limiting or prohibiting rather than taking direct action.

But when the State takes into its own hands what was previously left to private enterprise, then *the model of direct management* is involved. There the State nationalises private interests, sets up it own organisations and becomes an oil entrepreneur. Thus the state agency becomes a supplier of capital, a controller of activities, an employer of personnel, a buyer and seller seeking suppliers and customers, a signatory of industrial and commercial contracts. This is chronologically and logically the last stage of political intervention in the oil sector.

However, the creation and development of a state company does not alter the fact that the State has a more general rôle in legislation and control. It should be borne in mind that whatever advantages the State gains by entering the field as a private entrepreneur, an additional bonus is obtained by the mere exercise of its power to govern events; the economic activities are directed and controlled by political forces.

Accordingly, this assessment of oil politics will begin with the newer and more particular manifestations of state participation (the State as entrepreneur) and progress to the more general and collective effects (the State as arbiter). This chapter will discuss direct intervention by the State and succeeding chapters will describe the indirect intervention.

In speaking of the public sector I am referring to the behaviour of the State when it intervenes directly in oil industry and trade. That ultimate stage of government action can be reached in two ways, either by nationalisation of privately-held assets already existing or alternatively by setting up firms in which the State proposes to be the sole or majority shareholder. State authorities impose the change from private to public ownership or introduce a public sector when they consider private interests in the sphere to be unrepresented or weak or too dependent on centres of decision outside the host country.

The European consumer countries were the first to adopt the procedure but they usually refrained from taking over the facilities already held by private enterprise. More recently the producer states of North Africa have followed that lead but have often taken more drastic action: where they thought foreign interests were getting in the way they have not hesitated to sweep them aside. Aware of their initial disadvantage compared with foreign private or public organisations, they have chosen expropriation because it is a short cut and because of their lack of trained personnel and the know-how required for rapid progress in achieving an integrated system of purely national oil operations. Despite the differences in method, the essential object of the exercise for consumers and producers alike was to obtain a degree of national independence.

Now let us look at how state companies on either side of the Western Mediterranean have fared. Critical assessment should show whether or not the results satisfy the hopes proclaimed. From a study of the circumstances of the state companies and of their shortcomings it will be possible to determine more clearly the rationale for operating enterprises of this kind, before drawing conclusions on their chances of survival.

Circumstances of the state companies

Some specific aspects of the circumstances of the state companies of the Western Mediterranean coastal nations should be examined. How do results compare with intentions? Do the state companies really serve national policy? How big are they? Have they achieved complete integration of the various stages of oil operations?

Relationship with government policies

Whether they belong to producing countries or to consumer countries in the area under special consideration, the state companies are regarded as a special means of implementing government policies. But in practice are public enterprises always and everywhere mere tools in the hands of the political authorities? In fact, their degree of autonomy varies according to three factors – the amount of oil discovered, how long the company has been in existence and the ideology of the politicians. Hence the variety of the results, although certain general features can be seen in them.

In the producer countries, state dominance increases from Morocco to Tunisia, then Libya and finally Algeria. Morocco is interventionist in oil affairs but the effects are not very obvious because so little oil is being produced. In Tunisia the output is small and the State has joined in, through its holdings in partially state-owned companies exploring for and exploiting national oil resources. Libya has a state company, Noc, but this is comparatively recent and insignificant in a scene dominated by the big international groups and independents occupying the attention of the government authorities. Noc has preferred to seek joint endeavours with

French, Spanish or Italian public enterprises in trying to overcome this initial handicap but it has to make its way in competition with firmly entrenched forces. It came on the scene when matters had already been efficiently organised. The Libyan authorities profess the wish to put the state company into a strong position but it started almost from scratch and was not formally incorporated until 1968.

Sonatrach is in a different position, having taken over the publicly-owned interests in Algerian oil created during French trusteeship. The Algerian enterprise was the sole negotiator on hydrocarbons in the co-operation agreements with France made in July 1965. Its holdings have been considerably enlarged in consequence of successive steps taken by the government to 'recover' the large oil assets in 'foreign' hands. Of the many oil enterprises operating in the territory, Sonatrach is the only one that is trusted by the Algerian authorities. There is no comparable organisation in the other countries of North Africa. It is indeed a vehicle for the Algerian Government's socialist and national policies.

The conditions in which the state oil companies of the European countries operate are unlike those on the other side of the Mediterranean. ENI is the oldest and Hispanoil the most recent, with ERAP midway between. The organisational structure of Hispanoil resembles that of CFP. Both of them are partially state-owned companies, but in Hispanoil the State provides a large proportion of the directorship and the remainder comes from banks which also have heavy stakes in Spain's industrial development. In CFP the State's representation is small but arrangements have been made (government-appointed executives, shares with plural vote, control over important decisions) to ensure its preponderance and the remainder of the shares are spread over a multitude of unorganised holders.

These European companies have the common feature of being subject to competition from private enterprise. Although given favoured treatment by the Government, the state companies must make their way in a fairly competitive situation: successive administrations have not seen fit to give them a monopoly in their field. They have to work in relative harmony with the national networks of the big foreign groups.

Moreover, in France the competitive aspect is enhanced by the fact that state policy in the industry is channelled through two separate oil companies, ERAP and CFP. This situation is not found in Italy or in Spain, each of them having only one organisation representing direct intervention. The set-up in France provides greater scope for manoeuvre, because the possibilities of either of the two enterprises can be brought into action according to circumstances. This should be quite a good way of avoiding the bureaucracy or excessive concentration on political considerations which are likely to occur in a state monopoly.

There is thus variety in the type of facilities available for collaboration between state enterprise and the national governments. Another aspect which needs to be considered is the scale of public enterprise.

Size of public enterprises

State oil companies in the countries of North Africa other than Algeria are too small to engage the State in any sizeable activity in actually performing entrepreneurial operations. Among the European companies Hispanoil is still in an early stage of development, although it shows promise.

Sonatrach, ENI and ERAP may be regarded as true public enterprises.

The organisational arrangements of Sonatrach are still somewhat tangled, in that the main company engages in all kinds of activities which in other groups are assigned to various affiliates. Only the exploration work is becoming to some extent separated from the central unit, partly because of the setting up of partnerships with foreign interests, usually American. It is difficult to discover the true position regarding the consolidated accounts and change in the internal capital of the enterprise and hence to ascertain the turnover from all its activities. Sonatrach has ventured into the sphere of exploration which is a costly one. Also much of the crude in the company's hands has come from purchases, from seizure or from nationalisation of foreign companies. Moreover, it is impossible for an outside observer to distinguish clearly which acts have been performed by the Algerian State and which by its state company. At the same time the assets of the enterprise have frequently been changed by successive additions of nationalised or confiscated assets. Therefore it is difficult to give any definite figure of turnover. In 1970, before the 'recovery' of the French assets, it is likely to have been just over 1·5 thousand million francs.[1]

The position will probably become clearer in the course of time, with the general structure of Sonatrach resembling that of all the sizeable oil companies. This state company now acts alone in some operations (in exploration, production and selling) but it has grown too quickly to bring the organisation of its affairs into line with its present scale of activities.

ENI and ELF have adopted a classic plan of organisation; they have a central company responsible for promoting group policy and mobilising the resources for putting it into effect. The affiliates are interlinked around that central unit, which acts as a holding company. Operations are

ENI

AGIP

- AGIP SAUDI ARABIA
- AGIP TAILANDIA
- CORI (Libya)
- SOMICEM
- SARCIS
- AGIP UK
- AGIP (Netherlands)
- NORSK AGIP (Norway)
- AGIP RECHERCHES CONGO (Brazzaville)
- AGIP RECHERCHES ET EXPLOITATIONS PÉTROLIÈRES (Madagascar)
- NAOC (Nigeria)
- SAEP (Tunisia)
- SITEP (Tunisia)
- IMINOCO (Iran)
- SIRIP (Iran)

- AGIP CANADA
- AGIP PETROLERA ARGENTINA SAMIC-YF
- AGIP PETROLEUM (USA)
- IEOC (Egypt)
  - COPE (Egypt)
- AGIP PETROLEOS COLOMBIANOS
- ROMAGAS
- SERAM
- VADOIL
- AGIP SUISSE
  - Stockage
- AGIP AUSTRIA
- AGIP ERDÖLGEWINNUNG

- AGIP ESPAÑA
- AGIP (MUNICH)
  - Chepromin
  - Chepromin & Co. KG
  - Touring Tankstellen
  - Tanklager Hanau
  - Tanklager Hanau & Co. KG.
  - Neusser Tanklager
- AGIP HELLAS
- AGIP FRANÇAISE
- AGIP (Brazzaville)
- AGIP (Cameroons)
- AGIP CASABLANCA
- AGIP (Congo)
- AGIP (Ivory Coast)
  - Société d'Entreposage San Pedro

- AGIP (Dahomey)
- AGIP (Ethiopia)
- AGIP (GABOON)
- AGIP (Ghana)
- AGIP (Liberia)
- AGIP (Kenya)
- AGIP (Madagascar)
- AGIP (Nigeria)
- AGIP (Sierra Leone)
- AGIP (Somalia)
- AGIP (Sudan)
- AGIP TANZANIA
- AGIP (Togo)
- AGIP TUNIS

- AGIP (Uganda)
- AGIP (Zambia)
  - Nosco-'Ndola Oil Storage Co. (Zambia)
- PETROLIBIA
  - Asseil
- AGIP ARGENTINA
- AGIP (Cyprus)

- SEMI
- STEI
- SOCIETA AUTOSTRADE CENTRO PADANE
- SOCIETA PER L'AUTOSTRADA DELLA VAL SERIANA
- SIPO
- SOCIÉTÉ HOTELIÈRE SAMIR (Morocco)

NUOVO PIGNONE

- FUCINE MERIDIONALI
- PIGNONE INC. USA
- PIGNONE ENGINEERING (GB)
- PIGNONE SUD
  - PIGNONE SUD IBERICA
- SOCIÉTÉ INDUSTR. et COMM. PIGNONE FRANCE
- PIGNONE ESPAÑOLA

LANEROSSI

- LANEROSSI (W.Germany)
- LANEROSSI FRANCE
- LEBOLE EUROCONF
  - Gagliano Confezioni
  - LOBSTER
- MARLANE
- ROSSITEX

- SAPEL
- SOCIETA EUROPEA ROSSIFLOOR
- THERMOTEX
- ROSABEL

SNAM PROGETTI

- EQUIPGAS (Spain)
- SNAM AUXINI PROYECTOS
- SNAM INA PROJEKT
- SNAM PROGETTI AUSTRALIA
- SNAM PROGETTI FRANCE
- SNAM PROGETTI (Switzerland)
- SNAM PROGETTI USA

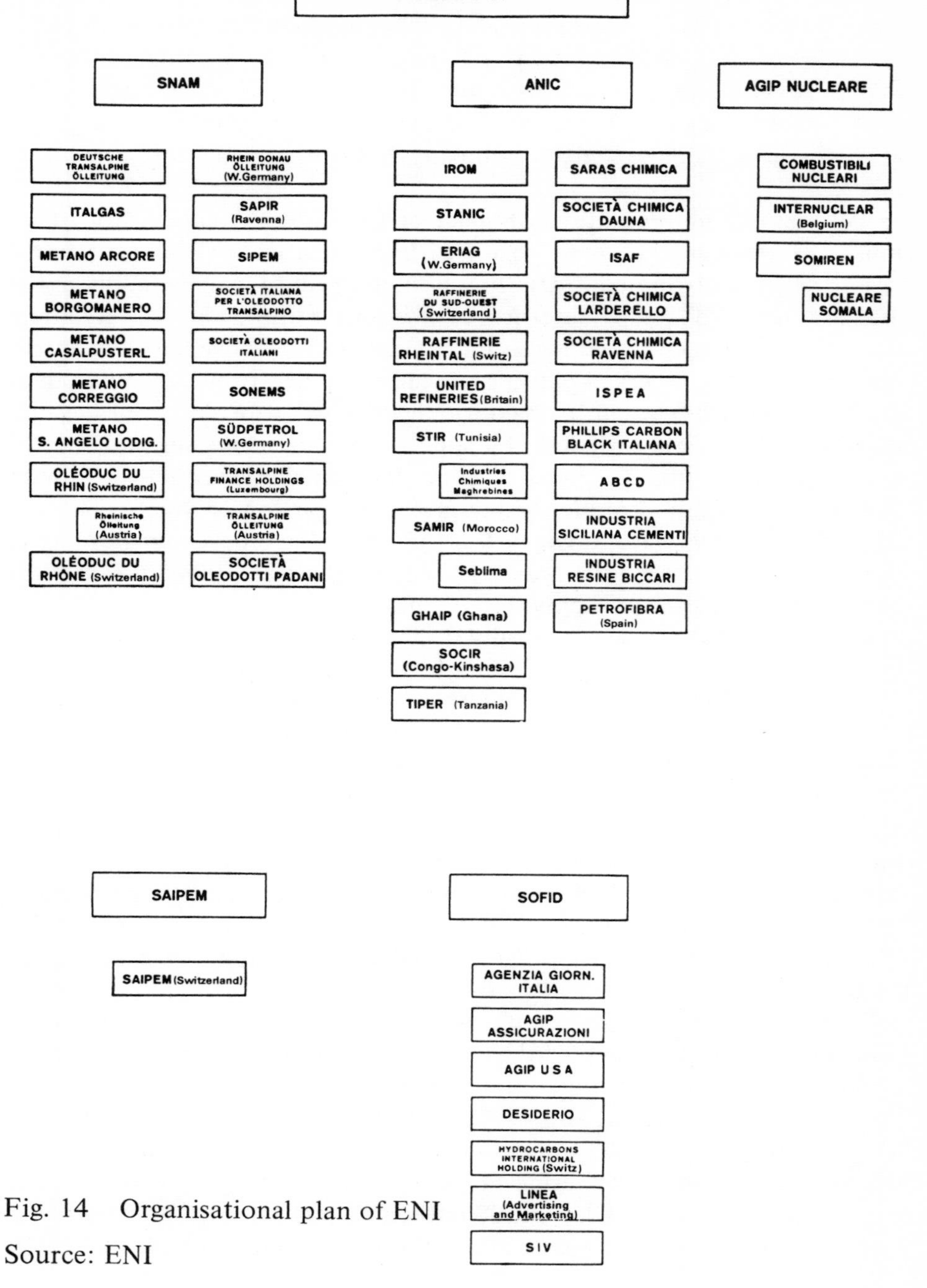

Fig. 14 Organisational plan of ENI
Source: ENI

divided up according to the type of activity performed or the region involved. Thus, for example, AGIP heads production and exploration for the whole complex but has set up local companies in the various countries in which it is operating. ELF RE is in a similar situation, directing exploration and exploitation of hydrocarbons in France and abroad. The consolidated balance sheets of ENI and of ELF only include those affiliates in which the central company has a holding of 50 per cent or more.

Turning to the international impact of the state companies, Sonatrach must be considered to have begun to have significance at international level. It is negotiating contracts for the sale of oil to its neighbours in the Maghreb and to some countries of Western Europe, and is gradually extending its exploration to certain Arab states, above all Yemen. In natural gas the Algerian enterprise will soon be firmly established at international level: in all probability the exports to France will develop normally, shipments to Britain by supertanker are increasing and the US authorities have finally agreed to allow the valuable contract made with El-Paso (15 thousand million cubic metres a year for 25 years). There is such a shortage of natural gas in Europe, as well as an even greater one in the United States, that the industrialised countries are unlikely to neglect any source of long-term supply.

The Algerian enterprise had only 3,000 employees at the end of 1967 but the figure rose rapidly to 9,000 in 1970[2] and 10,000 at the beginning of 1971, which is half the total number in ELF/ERAP. The figure may be expected to become very much higher because of the new tasks set for the company and because of the nature of unemployment in Algeria.

By comparison with the Algerian company, ENI and ELF are already well established at international level, although they differ from it in the proportions of turnover from the various types of activity. Because no significant quantities of oil have been found in the subsoil of France and Italy, exploration has necessarily been transferred to other countries. However, because of tariff situations and quota restrictions, refining has traditionally been carried out in these countries and the extension of oil activities to other countries only became possible later, when the initial investments proved to be profitable and served as a basis for expansion.

The turnover of those two public enterprises puts them in a creditable position in world trade. They are in the top fifty in the list of the 200 largest enterprises of the world outside the United States. According to the list published in the journal *Fortune,* ENI was in thirty-third place in 1969, with a turnover of 1,616,800 million dollars and 62,733 employees. Immediately above ENI was CFP, with a group turnover of 1,624,775 million dollars and 23,000 employees. ELF/ERAP came in forty-second place, with 1,274,131 million dollars and 16,455 employees.[3] This figure related to the ELF group as a whole and not ERAP alone. The latter had to dismiss 600 of its employees when its exploration and production operations in Algeria were curtailed.

The high place in this roll of honour achieved by ENI is impressive, although the ratio of employees to turnover is disquieting. From that point of view CFP and ERAP did better than the Italian enterprise, or for that matter other oil companies higher up the list.[4] Study of the profit and loss account of ENI for 1969 shows that from 1968 to 1969 labour costs produced the greatest rise in expenditure (+18 per cent).

Division of turnover by number of employees of the enterprise can give an indication of productivity per person. This calculation reveals that putting all the activities under the control of a single group is not necessarily the best way of conducting oil business. Whilst this can, of course, result in savings, an unwieldy apparatus may develop, with too many main offices, duplication of work, a proliferation of resources all over the world and reduced profitability of the whole. ENI seems to have contracted that disease, to which public enterprises are more prone than private ones. Increase in the payroll should be accompanied by constant rationalisation of operations, which requires firm management.

There is, of course, an element of risk in some branches of oil operations. Even well-planned exploration may yield nothing but loss for year after year. ENI experienced this in Morocco. On the other hand, in Libya, Occidental was lucky enough to discover an exceptionally rich deposit within a few years. And below a certain level refining and distribution may not be very profitable, especially if there is strong competition in those lines. In those respects ENI seems to be in a more favourable position than ELF. However, the addition of ANTAR will probably improve ELF's results. The ANTAR refineries are mainly in Western and Northern France, whilst those of ELF are for the most part in the South West, the alpine region around the Rhône and the Paris area. ANTAR has a refining capacity of 10 million metric tons, possesses 6,000 sales outlets, has only 4,000 employees and holds a 10 per cent share of the domestic market.[5] Before the takeover it was a well-managed enterprise showing a financial profit from refining and distribution because of efficient organisation. It will still be able to obtain supplies of crude from CFP, as recent agreements give the latter a stake in the capital of the enterprise.

Part of the strength of the ELF Group comes from its association with SNPA. The consolidated turnover of that organisation for 1969 was 1,480

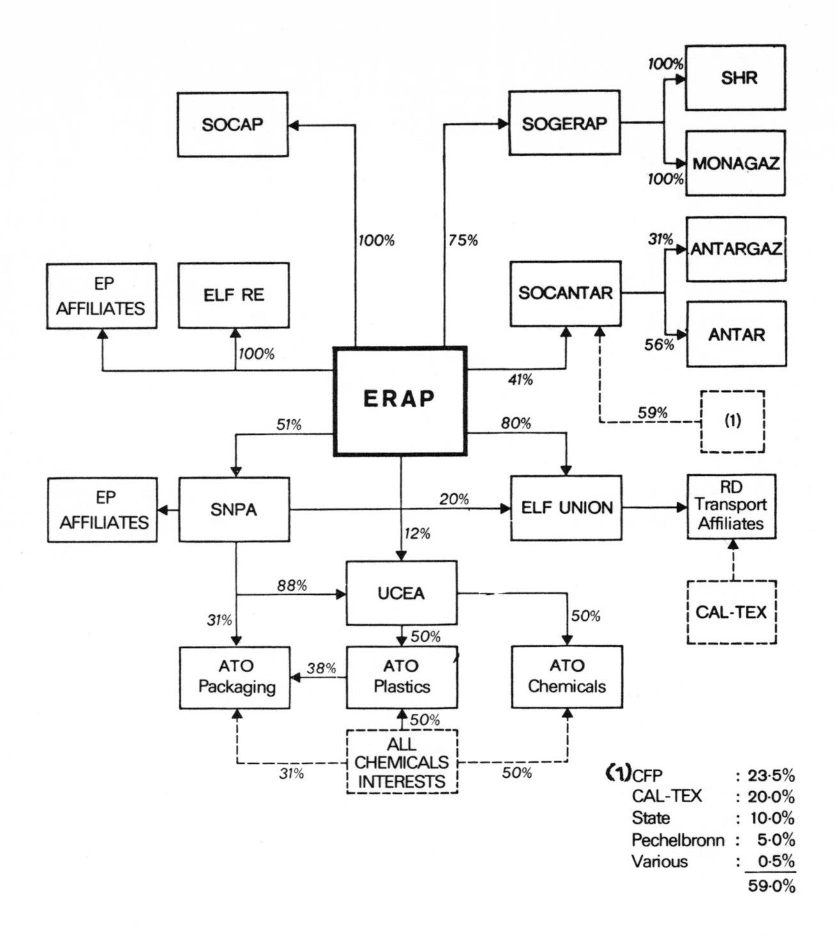

Fig. 15 Structure of ELF group, July 1972

Source: ELF

million francs, i.e. nearly a quarter of the total for the Group (6·3 thousand million francs). But the deposits at Lacq are rapidly becoming exhausted and the sulphur content of the oil will henceforth greatly affect profits – and oil from that source accounted for 30 per cent of SNPA's turnover in the early 1970s. On the other hand, Aquitaine's petroleum products have benefited from the favourable market for them since 1968 and they have also done well in Canada. SNPA's profits for 1969 were 340,095,000 francs, whilst those of ELF were only slightly higher, at 372 million francs.[6] The merger of ELF and SNPA is to the detriment of the funds of the latter, since SNPA's profits have been diverted to finance the group as a whole. SNPA's vigorous growth would need more capital in order to continue at the rate that has been achieved, but the extra capital is not available because ELF would have difficulty in finding the funds to increase its holding accordingly. Alternatively its representation would be reduced if it did not contribute to an increase in the capital of SNPA.

For ENI, undeniably the situation is not healthy, despite the undoubted expansion of the company. Since 1953 the net profit of this state company has shown moderate improvement; in 1967 it was 28·8 million francs (3·6 thousand million lire). It was very little better in 1968, when it reached 31·2 million francs (3·9 thousand million lire).[7] Admittedly the situation showed a definite improvement in 1969, when the profits were up by 169·2 per cent to 84 million francs (10·5 thousand million lire).[8] However, they are still small for the scale of ENI's operations.

The thriving sector of ENI, comparable to the part played in ELF by SNPA, is pumping, transport and sale of methane in the wealthy industrial Po Valley. The amount of gas distributed in 1968 was 10·3 thousand million cubic metres, showing a 15·5 per cent increase over 1967. But the deposits in the North are becoming exhausted and ENI is being obliged to turn to much more costly off-shore operations in the Adriatic. The problem for ENI, as for other public enterprises, is that it has difficulty in finding funds to match its desire to take part in all stages of petroleum operations.

Degree of integration

All the public oil enterprises of the Western Mediterranean coastal states strive for complete integration of control of the oil from the moment that it flows from the well, endeavouring to obtain sole possession of it throughout the various stages of processing and handling until it reaches the user. When drilling in distant countries is undertaken, efforts are being made to secure sole management of the brand from the well to the final consumer. First Sonatrach and then NOC, also ELF, ENI and later Hispanoil, have tried to emulate the majors by taking part in the whole range of oil operations – exploration, production, transport, refining and distribution. They consider it indispensable to their business success to possess all the links in the chain.

It must be pointed out that the public companies are still a long way short of the target. Despite their efforts, some of the links in the chain still look very fragile.

Sonatrach, for example, is now in a dominant position in production, accounting for two-thirds of Algerian oil output. It also has sole control of the flow of its oil through the pipelines in Algeria. But upstream and downstream on either side of those operations it still has very few facilities. It refines and sells finished products almost solely for Algerian consumption, which is as yet small. A vast area for exploration is available to it in the Sahara but its financial, technical and human resources are still relatively poor. Nevertheless, by 1969, when the French were still exploring the subsoil of the Sahara region, there had already been considerable investment in the search for oil: 280 million dinars, constituting a little over 50 per cent of the sums that had been spent on such operations in the Sahara.

ENI has better results from refining, re-export of processed products and distribution than from production or exploration. The reasons for this curious bias lie in the two strategies always applied simultaneously by Mattei and his successors. Firstly, there was a wish to be on friendly terms with the new independent republics which have arisen from the vast post-war decolonialisation movement, above all those of Africa. The Italian strategists quickly saw the implications of that redistribution of political power. They offered to set up refineries in the new states and to give the host government a holding in the venture. This occurred at a time when the majors were unwilling to adopt such an approach, preferring to supply the newly independent states from larger refineries located in the developed countries, in order to make savings from the scale of processing operations and to take advantage of the possibility of balancing supply and demand from several large markets. In taking this attitude the majors certainly under-estimated the many small benefits which ENI later derived from its policy, which were of some assistance to it when competition was fiercer. ENI thus managed to obtain for itself a multitude of sales outlets, spread throughout the world, for small quantities of crude.

The other ENI strategy departed from the traditional chain of oil operations and sought to use differing sources of energy (oil, natural gas and now uranium) to create an industrial complex with which Italian products could more easily penetrate foreign markets.

ENI has always had difficulties in connection with the production part of oil operations. Mattei was described in international business circles as an 'oilman without oil'.[10] His successors still suffer from the same misfortune. Substantial deposits have been discovered but although the company's news bulletins keep announcing imminent exploitation, this is constantly postponed from one year to the next. There have been difficulties because of the insecurity of the situation in Nigeria, where the enterprise has, in fact, made worthwhile discoveries. There are further specifically political problems in connection with the deposits in the Sinai Desert, where production began shortly before the 1967 war; Egypt was then the host state but the area came under Israeli occupation. In the case of the very promising Concessions 100 and 101 in Libya, the obstacles are above all technical – the crude is mixed with paraffin and is difficult to extract. Although good levels of output have been achieved in Tunisia, Iran and Saudi Arabia, the company's total production has been very slow to increase. In 1968 it was 7 million metric tons and in 1969 it was 8·1 million. Even with the encouraging results of operations in the North Sea and in certain Persian Gulf emirates, in 1972 the enterprise fell short of the '20 million tons a year' which it had hoped to attain.[11]

Efforts have been made to extend the area of the group's concessions. At the end of 1968 the total area of the company's concessions, after deduction of shares belonging to partners in joint ventures, was 381,612 square kilometres; in June 1970 ENI's net area was 100 million acres, equivalent to 450,000 square kilometres.[12]

By contrast, the ELF + Aquitaine enterprise had a net area of 1,261,157 square kilometres on 31 December 1969, compared with 1,263,000 square kilometres in 1968. The French group had markedly better results than ENI in the production of crude, with a figure of 24·4 million metric tons in 1969. But the volume of production was insecure, in that more than 19 million tons of that total came from Algeria at a time when French interests in that country were threatened. A vigorous plan of diversification of the sources of supply was launched in 1960, which was already rather a tardy move. However, there was a delay in imposing expropriation in Algeria. In 1972 it was considered necessary to double the development programme for the deposits already discovered, from 450 to 900 million francs. The nationalisation measures adopted by the Algerian Government in spring 1971 dealt a heavy blow to the French company. In 1972 the area of ELF's concessions was about 10 per cent smaller. In that year Aquitaine is reported to have regained its 1970 level of production (2·6 million metric tons, including slightly more than 1 million in Algeria). However, for ELF it was more difficult to restore the volume. There was a net drop in production in 1971, in consequence of the elimination of two-thirds of the output in Algeria. The production figure was 13 million metric tons, including 6 million from Algeria and 4 million extracted in Gaboon. According to information supplied by the company, the 1970 level of production might be restored in 1975, through the commencement of exploitation of the deposits discovered in 1972 in the Ekofisk field in the North Sea and in the Emerald field in Congo, as well as by resumption of exploitation of the Obagi field in

Nigeria, in partnership with the Nigerian state company, and very likely from further discoveries in the large total area of concessions prudently distributed over various parts of the world other than the closely guarded preserves of the Middle East.[14]

It is interesting to note that in 1968 the ELF group was making a greater effort than ENI in 'exploration and development'. The consolidated balance sheet showed 1,207 million francs, 66 per cent, out of total investments of 1,825 million francs, for 'exploration and development', 19 per cent for refining and distribution, 7 per cent for petrochemicals, 1 per cent for natural gas and 7 per cent for general and sundry investments.[15]

In the same year ENI spent 808·8 million francs on 'prospecting and extraction', which represents 39·3 per cent of its total investments (2,059·2 million francs), whilst 19·9 per cent was spent on transporting and distributing methane, 24·8 per cent on refining and distribution and 4·8 per cent on the chemicals sector.

These figures illustrate differences in emphasis in the policies of the two state companies, the French one placing more stress on exploration and the Italian one concentrating especially on continued investment in methane supply and on investment in refining and distribution. Both companies keep expenditure on all aspects of the petrochemicals branch small, because their decisive efforts must be made in other operations. ELF, for example, expected to lose a high proportion of its oil interests in the Sahara and was engaged in a desperate effort to survive, or at least to prevent loss of autonomy. The group had to force the pace elsewhere, above all in the franc zone (Gaboon and Congo).

The Italian group, for its part, cannot simply resign itself to its inherent weakness in ability to supply crude bearing its own brand name. It wants, at all costs, to obtain 'a sufficient degree of autonomy' in its internal supplies. ENI seems to have decided to design its policy accordingly in future and to increase the proportion of investment in upstream operations. For the period 1969–72 the group planned to invest 7,600 million francs (1,440 in Italy and 6,160 abroad), representing 45·9 per cent of its total investments, on searching for and producing hydrocarbons.[17]

In all these companies transport is undoubtedly still a weak link in the chain of integrated operations.

Sonatrach, ENI and ELF possess good pipeline systems, which are an important source of profit. But everywhere, even in their own national territory, the state companies are obliged to negotiate with other powerful interests in the business over transport arrangements.

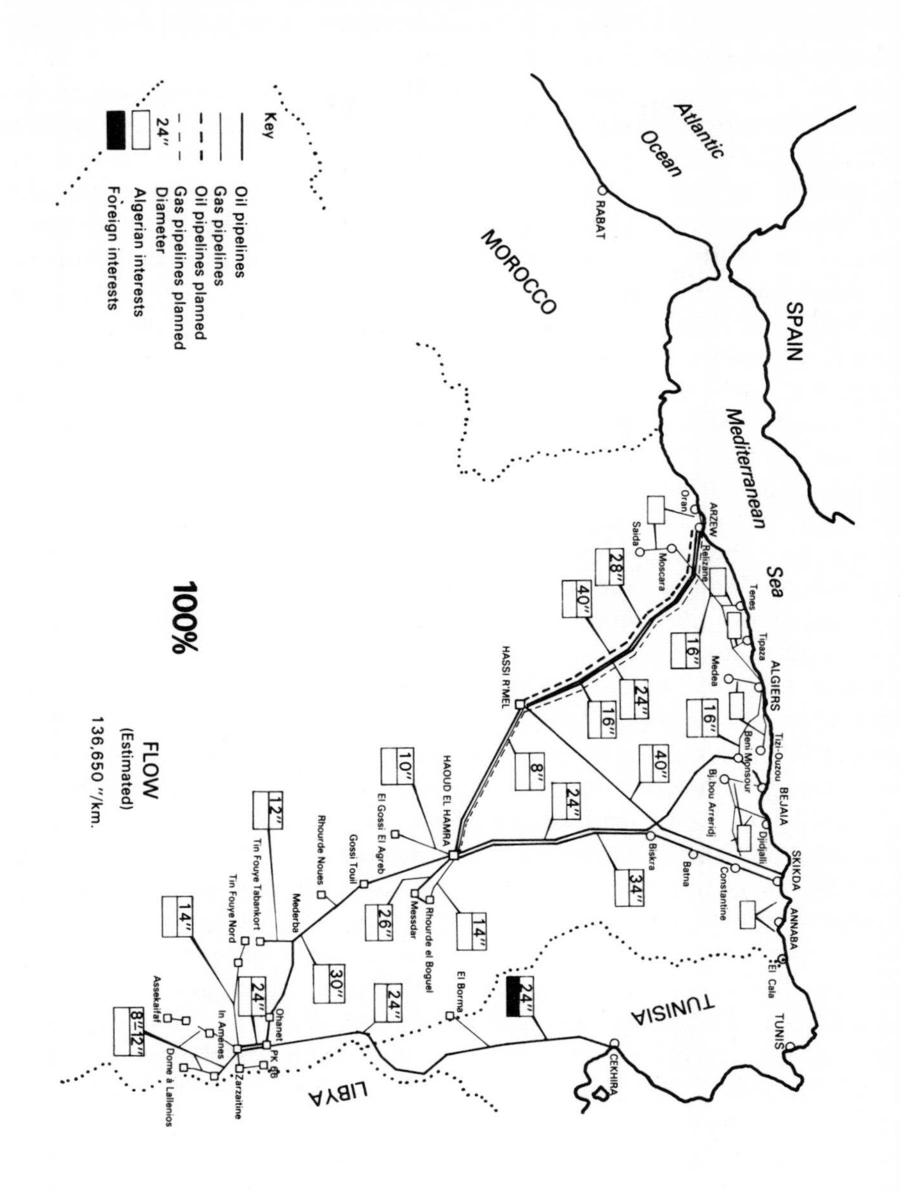

Fig. 16 Hydrocarbons transport in Algeria, 1972

Source: *Revue Française de l'Energie,* 255, July–August 1973, p. 412

It will take the Algerians a long time to forget the experience of discovering that all transport in their territory was in the hands of foreigners. By 1969 Sonatrach had a half-share in the Algerian pipelines and it has since acquired full ownership.

In the European states with Mediterranean coastlines there is great rivalry between Marseilles, Genoa and Trieste to capture oil from North Africa or the Eastern Mediterranean ports serving as terminals for the Middle East pipelines. The ports have been vying with each other to be best able to supply Bavaria and the Rhine area. This competition is a good illustration of the general shift from national insularity and from the exclusive claims connected with it. The system of transporting oil through pipelines which begin in Southern Europe is largely organised by the majors, who draw up their plans from long-term forecasts of demand in which the whole of Western Europe is treated as one planning unit. In the TAL pipeline running from Trieste to Ingolstadt, ENI is only allowed a 10 per cent share. ENI is the sole owner of the pipeline from Genoa to Ingolstadt and to Aigle, in Switzerland, but the flow in the early 1970s was only 8 million metric tons a year. For any improvement in profitability the state companies have to rely on an increased volume of flow through the pipelines rather than on an enlargement of their holdings in the pipeline consortiums, as table 3 shows.

The aggregate capacity of Marseilles, Genoa and Trieste for handling oil could be raised to 152 million metric tons during the 1970s. But of these three ports Marseilles is likely to receive the greater part of the increase, because the Fos-sur-Mer harbour has the best facilities for supertankers. In 1970 there were plans for raising the capacity of the pipeline from Fos-sur-Mer to Karlsruhe from 34 million metric tons to 90 million metric tons. In 1972 the capacity was increased to 42 million metric tons and there was a possibility of further development raising the flow to 65 million metric tons for 1973. The ELF group refineries near the route, above all at Feyzin, in the suburbs of Lyons, were taking more than 9 per cent of the flow. After various negotiations and after the French authorities had put pressure on the bigger companies involved, the French state company had been able to obtain a larger share of the volume carried by the pipeline.[18] The interest in the consortium which ANTAR already held before its merger with ELF is a valuable addition.

The state companies seem to be severely handicapped by a shortage of shipping, especially by comparison with some of the majors which are well-equipped in that respect, for example Shell.

The Algerian authorities adopted an order of priority and first tried to obtain expansion and control of pipeline facilities in their territory. They

Table 3

Holdings in pipeline consortiums at beginning of 1970
(percentage of total)

	South European pipeline (PLSE)	Central European pipeline (CEL)	Trans–alpine pipeline (TAL)	Rhine–Danube pipeline (RDO)	Adria–Vienna pipeline (AWP)
Esso	28·0	–	20	20	6·5
Shell	23·2	–	15	15	14·5
CFP/CFR	15·0	–	2	2	4·0
BP	8·0	–	11	11	7·5
Mobil	5·6	–	11	11	12·5
Texaco/DEA	4·0	–	9	9	–
ANTAR	4·0	–	–	–	–
Gelsenberg	3·6	–	6	6	–
ELF	2·8	–	–	–	–
Veba Chemie	2·4	–	3	3	–
Wintershall	2·4	–	3	3	–
Petrofina	1·0	–	–	–	–
ENI	–	100	10	10	4·0
Marathon	–	–	7	7	–
Continental	–	–	3	3	–
OMV	–	–	–	–	51·0
	100	100	100	100	100

PLSE runs from Fos to Karlsruhe. Flow = 34·4 million tons p.a.
CEL from Genoa to Ingolstadt and Aigle. Flow = 8 Mt p.a.
TAL from Trieste to Ingolstadt. Flow = 25 Mt p.a.
RDO from Karlsruhe to Ingolstadt. Flow = 12 Mt p.a.
AWP from Adria to Vienna. Flow = 6 Mt p.a.
Esso, Shell and BP have an agreement for a flow of 4 million tons a year for 20 years through CEL.

See figure 13, pp. 76–77

then manifested a wish to control the shipment of oil also. However, progress in obtaining the shipping has been slow. It can only be achieved in close co-operation with France. The French Government did, in fact, undertake in 1965 to give Algeria the technical assistance required in order to be able to operate a national fleet of oil tankers. It also promised to do its best to get French importers of Algerian crude to charter the tankers built in the French naval shipyards, provided the terms and conditions of shipment were competitive. The original plan was for Sonatrach to have a fleet of three tankers of 60,000 tons. [19] Apart from the fact that the 1971 crisis upset the arrangements for co-operation, the emphasis had already shifted from oil to natural gas. International markets are open to the fantastic resources of Hassi-R'mel. This gas has to be transported in liquid form at very low temperature. France has acquired some expertise in building methane tankers. In 1970 the French naval shipyards delivered a 40,000 cubic metre third generation methane tanker to the owner, Compagnie Nationale Algérienne de Navigation, a state enterprise. Vessels of 125,000 cubic metres are now being built, at a unit cost of 500 million francs. The La Seyne shipyards, in France, are due to deliver two of these to Algeria by 1978.

Since oil became an important commodity the Governments of France, Spain and Italy have tried to redress the shortage of shipping. A French decree of 19 March 1938 obliged importers to ship not less than 50 per cent of the tonnages of crude purchased in vessels sailing under the French flag. In 1950 decrees concerning the renewal of special import permits raised to two-thirds the proportion of shipment to be made under the French flag.[20]

The Hispanoil fleet has only been carrying a small tonnage but Government pressure is likely to enforce a rapid increase. In 1968 ENI spent only 24·8 million francs, amounting to 1·2 per cent of its investments, on building a fleet of its own. [21] In the same year SNAM tankers carried only 3·8 million tons of crude and petroleum products.[22]

Crude oil shipments by the ELF group in 1968 were more than 16 million metric tons but 67 per cent of this amount was carried in vessels on time charter, [23] 16 per cent by vessels chartered for consecutive sailings and 17 per cent by vessels chartered for one voyage. [24] State help has been invaluable in allowing the group to avoid an excessively heavy burden of investment. The situation of having to obtain shipment by third parties is not too unacceptable when exchange rates are stable. However, after the closure of the Suez Canal in 1967 the demand for large tankers was greatly increased in order to reduce costs per ton for oil shipped by the Cape route. 'On 1 January 1970 the tonnage of oil shipping under construction or on order in the world was equivalent to more than 50 per cent of the fleet in service'. [25] This new demand and the substantial rise in insurance premiums, because of several disasters involving pollution of beaches or fires, together caused a large rise in freight rates in 1970–71. It has been calculated that shipping costs of crude for the European markets in those years were as high as refining costs. But the cost of shipment for a long distance is greatly reduced by increasing the volume carried per vessel. Shipping costs in the first half of the 1970s were high for state companies which did not have sufficient tonnage in vessels of their own or were not in a position to invest in supertankers. But groups with capacity in excess of their own requirements, among them CFP, or which had both the foresight and the capital to get a fleet of supertankers built very quickly after the closure of the Suez Canal, as Shell did, hardly suffered from the shipping situation or even profited from it. However, the position in regard to the availability of tankers fluctuates wildly and after a difficult eighteen months supply and demand in relation to shipping space came into line in 1972. The market suddenly settled and artificially high prices could not continue to prevail.

From the factual details given above it appears that of the relevant state companies of consumer countries only ELF and ENI can compare with the big oil companies. In the producer countries under review Sonatrach has managed to lift itself to a level of international importance but exact figures for this recent achievement are not yet obtainable. With Government support and assistance these three state companies have become entrepreneurs of consequence in oil affairs at home and abroad. But in order to maintain or improve their positions despite world competition they must make good use of factors advantageous to them and eliminate some of their worst weaknesses.

Weaknesses of the state oil companies

Considerations of how and why the state oil companies were established have already been discussed and need no further explanation here. They derive strength from their inherently political nature, through government support, from links with the officials responsible for applying Government regulations, from boldness in their dealings with producer countries, from carry-over effect benefiting national industry as a whole if there is any lasting political harmony between two countries and from priority in the national market and in contracts from other state organisations (the Army, transport enterprises, public service undertakings).

As an illustration of this kind of strength, Mattei undoubtedly adopted an unprecedented line of approach in his initial commercial dealings with new countries emerging from colonial rule. He offered to sign contracts giving better terms than the fifty-fifty profit-sharing arrangements which had become the rule and which were being treated as an unalterable principle at that time. The signature of supply contracts on terms favourable to the supplier implies obtaining advantages in return. When Italy finally decided to import large quantities of Soviet oil the transaction was on barter terms involving the supply of steel tubes and pipes to the USSR in exchange for the oil. Whereas the commercial dealings of private enterprises are subject to some inhibiting factors, the dealings of public enterprises in an atmosphere of goodwill between countries may be much bolder.

Moreover, it is the function of state companies to take care of specific matters which are of importance to the nation and which might otherwise be ignored. These include security of supplies, competition in the market and hence better prospects for the interests of consumers, stimulation of private enterprise by challenging its hegemony, reduction of the trade gap and possibilities for better balance of payments situations when the national enterprise becomes capable of exports or re-exports.

However, it is unwise to take too optimistic a view in present day circumstances. The situation in Europe immediately after the ravages of the Second World War favoured state companies, but now governments may tend to help all companies, whether public or private, in important sectors. A responsible government cannot allow serious deficiencies to develop in key sectors of the economy. It will intervene and give assistance if necessary, whatever the ownership of the enterprise, as in the iron and steel industry in France. Professor Galbraith has pointed out the close liaison which exists between senior officials of the Administration and the top levels of American corporations, through what he calls 'the technostructure'. Does the same apply on the other side of the Atlantic? In Algeria, where there is a tendency to distrust foreign capital and national private capital holdings are small, the authorities show little sign of such goodwill towards private enterprise. In France, however, a trend in that direction began to appear after M. Pompidou came to power. Economic intervention by the State more readily affects private enterprise in Italy, under the 'planned contracting' system.

In present circumstances public enterprise is called upon to justify its existence. The mixture of politics and economics involved in state companies does not necessarily produce the efficiency which is so important nowadays. Many of the weaknesses of state companies arise from the very features which are their strengths from other points of view. This may be seen clearly when political considerations restrict business operations too much, preventing sound economic practice.

Some of the weaknesses to which state companies are prone will be discussed below. In fact, these arise not so much from intrinsic defects as from a tendency to develop certain faults in varying degree according to local conditions and the general situation. Five possibilities seem to be especially noteworthy, namely the danger of inflexibility, the danger of evasion of democratic control, the danger of ambiguity in the status of a public enterprise, the danger of indiscipline in financial affairs and the danger of political impediments at international level.

Danger of inflexibility

It is well-known that bureaucracy increases as competitive pressure declines. If a state enterprise has a real monopoly the absence of the spur of competition will very likely cause it to drift into complacency and laziness. It may even develop aims of its own and forget the purposes for which it was brought into being. In countries where there is a high proportion of unemployment, a state enterprise may become a kind of national workshop. In some circumstances the state company may shed its economic objectives and become a mere means of indulging the whims of politicians, thereby developing ostentation, extravagance, duplication of work and excessive bureaucracy.

Something of this kind has happened in Spain's Campsa, which in 1927 was granted a monopoly in importing crude oil. The company has neglected the function for which it was set up, falling into inertia because of lack of competition in the domestic market.

Campsa was formed in order to ensure state monopoly but is linked through the banking system with enterprises of all kinds, 48 per cent of the capital invested in the oil industry being provided by five large banks. [26] Campsa itself has drifted increasingly into the rôle of an intermediary, dealing with a mass of sub-agents who operate petrol stations, transport companies and refineries. The distribution network in Spain was still well organised until a few years ago but it has since become rather inefficient and in recent years refining capacities have been far below the country's needs. The cure for these shortcomings might be to put an end to this confused situation. Either the private operators might be officially recognised as serving the national economy at their level or alternatively Campsa might be given a new management team of go-ahead businessmen subject to subsequent scrutiny of operations by the authorities and working for the profitability of an enterprise provided with larger funds that are less dependent on indirect taxation. It would be necessary for their work at the head of a large rehabilitated public company to be performed openly, allowing the press and bodies representing the nation as a whole to see what was happening.[27]

There are risks of Sonatrach developing failings of the same kind as a result of those spheres in which it gains a definite monopoly and becomes the sole representative of Algerian interests. Undoubtedly prestige operations, inflexibility and some cost inflation occur in a system subject only to self-discipline. Sonatrach may become increasingly detached from the rest of the nation, devising and carrying out its own policy. This is a situation which can lead to high salaries in a poor country, creation of posts, adoption by an enterprise of a line of its own in dealings with foreign companies and the assumption of some independence in financial matters. In order to evade serious consequences a state enterprise may report its activities to the authorities after the event and endeavour to obtain approval by top officials of its own strategies.

Evidence that some practices exist in Algeria may be found in the Government Order containing the Finance Regulations for 1970, published on 31 December 1969.[28] The way in which requirements are

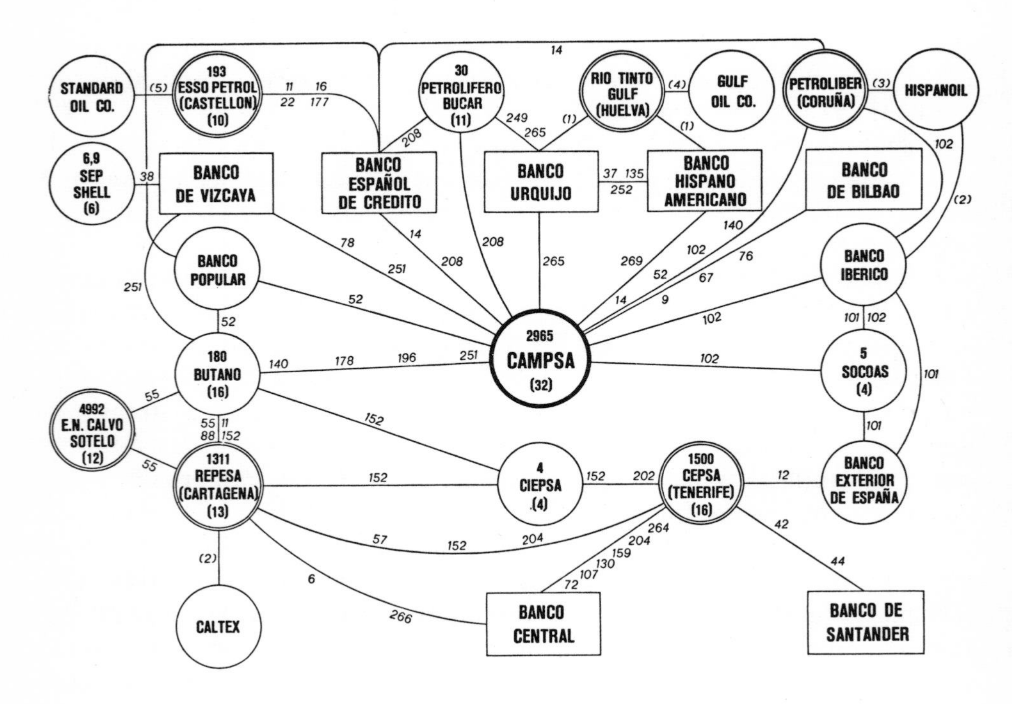

Fig. 17 Spain: links in the hydrocarbons field between enterprises

Source: R. Tamames, *Los monopolios en España,* p. 123

firmly emphasised in the text of those Regulations points to autonomous acts by the Algerian state enterprises and in particular by the most powerful of these, Sonatrach. There was some ferocity in references to the way in which financial affairs are conducted: for example, the Regulations demanded that 'the means of financing public investment in industrial operations planned or in progress as at 31 December 1969' should be revised in 1970 'to comply with the principles for finance stipulated under the Four Year Plan for 1970–73'.[29] Moreover, it was stipulated that state enterprises must each year submit to the Minister responsible for Finance and the National Plan their proposals for the following year, including 'in addition to their estimates of operating results and budgets . . . a statement of proposed means of financing production together with details of their production programme'.[30] From the 1970 financial year onwards, the Minister responsible for Finance and the National Plan was authorised to grant a loan to balance deficit resulting from the operations of state enterprises, but at the same time it was stated that in making such a loan there would be differentiation between 'deficits caused solely by factors external to the enterprise and imposed on it by the State' and 'deficits due to shortcomings in the management of the enterprise'. In the latter case the loan would be subject to acceptance of 'a plan for improving the working of the enterprise set by the Minister responsible'. This plan must stipulate above all 'the period within which the enterprise must put its management in order and the measures specified for such purpose'.[31]

Under further provisions for stricter control over the state enterprises the Minister responsible for Finance and the National Plan was required to appoint auditors 'for the purpose of ensuring that their accounts present a true and correct reflection of the situation and of investigating the position in connection with assets and liabilities'.[32] There was even provision for 'proceedings' against heads of these enterprises who did not submit by the required date 'their budgets and provisional accounts enabling the latest financial year to be provisionally closed'.[33]

Behind the formal language of these Finance Regulations there was clearly an attempt by orthodox financial experts to put the state companies on a responsible and orderly course of action as practised on the other side of the Mediterranean. Mr Abdesselam, the Minister responsible for hydrocarbons and de facto head of Sonatrach, felt that these provisions imposed excessive restriction of his freedom for manoeuvre. At the time of publication of the Regulations his resignation as Minister of Industry and Energy had been rumoured. However, he resumed charge of affairs, apparently having made up his mind to preserve the autonomy of his enterprise and refuse to give in to orthodox designers of financial policy who wished to bring state companies under state control.

The conflict was a true reflection of a political struggle between rival groups for the exercise of power in Algiers. For the time being the managers of the state enterprises prevailed over the controllers of national finance, thus postponing the time when a clear picture of the assets and liabilities of the largest state undertaking would be obtainable.

The natural inclination of state companies if they are not carefully watched by the nation's highest authorities is to depart increasingly from requirements for profitability. If the climate is too political there is a risk of the senior executives of those companies being appointed solely on the basis of political criteria, without regard to the needs of the company and good business management of its affairs.

Danger of evasion of democratic control

State companies are instruments of economic policies of Governments: in that connection they receive special subsidies and ought to be concerned to keep public opinion informed about their endeavours, because taxpayers certainly contribute part of the finance supplied to those enterprises. However, state companies often work in a semi-secrecy which is scarcely likely to gain them favour with the public, least of all among those who are in principle opposed to such organisations.

Algerians are aware of Sonatrach but they have little knowledge of the progress of its affairs. ELF and ENI do little to provide the public with the information to which it is entitled by virtue of those companies being public enterprises, but they do publish some material at times when pressure from public opinion may help them to overcome specific difficulties.

ENI still retains the habit of concealing its affairs to some extent and this doubtless derives from painful memories of the climate of hostility at home and suspicion abroad in which it initially had to make its way. In other countries there is little awareness of the virulent attacks to which ENI has been subjected from the beginning of its existence. A rapid review of the facts of that case will illustrate how difficult it can be for mutual trust to exist between public opinion and the managements of state companies.

At a very early date ENI was attacked by the Milan newspaper *Il Globo,* which specialises in financial matters. The newspaper alleged bad management, complaining that: 'In one of the regions richest in hydrocarbons not a single worthwhile oil deposit has been discovered. Exploration proceeds so slowly that there is a risk of its causing the irrevocable loss of national resources which could completely transform our economy.' The article questioned the need for keeping in existence a public enterprise with such a poor record.

From the beginning Mattei was opposed by the advocates of free enterprise. But there were also enemies abroad at a time when economic rapprochement between Italy and the United States was closest. He was simultaneously caricatured as David fighting Goliath and as St George slaying the dragon, that is, as the small person facing the giant and as an agent supported by all the might of the State in a struggle on behalf of the rights of the people. His rapid rise was regarded as being likely to impede the operations of Standard Oil, which had been firmly established in Italy since before the war, largely through STANIC. A clash in this connection was inevitable because the Italian Government had close ties with Washington under the Marshall Plan and NATO. The USA had a woman Ambassador in Rome after the war, Mrs Luce, who suggested that American loans might be reduced if Mattei's rise was not checked. At the same time the magazine *Time,* owned by Mrs Luce's husband, contained the statement that: 'In the eyes of American businessmen trying to create good relations with the Peninsula, Mr Enrico Mattei is the best, or clearest, example of what is wrong with the Italian economy'.[34]

Ample evidence is available to show the hostility to Mattei which persisted throughout his colourful career as head of the Italian hydrocarbon enterprises. There was a constant campaign against him in both the Italian and the foreign press. There was violent criticism in the British and the American news media. But the most vicious attacks of all came from the German-language Swiss press. The French press kept remarkably quiet, as if it did not attach importance to the phenomenon.

The press articles on Mattei and on ENI have been collected together in a set of volumes under the title *Stampa e oro nero* (The press and black gold). There are altogether 35 volumes documenting the press campaign conducted against the public enterprise and its head from June 1949, when the first discoveries of hydrocarbons in Italy were made, until 27 October 1962, the day of the air crash in which Enrico Mattei was killed.[35] For 1962 there are two volumes supplying a wealth of information fascinating to the researcher. It is possible to follow closely the reactions to Mattei's moves and the fear inspired in rational people by the boldness and speed of his actions. Here, for example, is the record of journalistic expression of the fierce opposition aroused by his project for a pipeline to Switzerland and Germany. Mattei made some fairly spectacular mistakes in that affair but his intuition was right. He realised that Europe ought to build a network of pipelines as soon as possible but he committed the error to which he was so prone of plunging into an operation without first investigating its economic viability. When he forced the hand of the big companies and upset their interests in this manner he then feigned astonishment at their furious reaction and complained bitterly of their disregard for the interests of host countries and their haughty 'imperialism'.

Another point which emerges clearly from these press articles is the tremendous hostility to Mattei's policy towards the USSR when the climate of the Cold War prevailed. At that time diplomats regarded negotiating with Moscow as dealing with the devil himself. Those who were less neurotic about the Soviet Union saw Mattei's activities abroad as acts of international piracy. Nevertheless Mattei managed to obtain from his negotiations with the USSR the result that throughout the agreement between three and four million tons of petrol was supplied at a price some 20 per cent lower than that which he was quoted by the big international companies. The policy of rapprochement with the USSR had two advantages. One of these was that of bringing pressure to bear on sellers in the West to lower their prices and thus cause sizeable price reductions by the companies which threatened his commercial independence: some competition was accordingly introduced into an oligopolistic market. Another effect was that Mattei's initiative proved rewarding in that it brought Italian industry a flood of orders, because in return for the crude oil Italy supplied products to the USSR, which had an urgent need for them because of the general embargo to which it was being subjecged. Soviet oil is still being paid for by the supply of Italian goods in a kind of large-scale barter arrangement. This amounted to selling Italian labour input in return for a source of energy which had to be procured somehow. The 1970 contract provided that 60 per cent of the cost of crude from the Communist bloc was to be represented by the value of Italian products supplied.

By far the most vitriolic attacks – and the most unjust – came from Montanelli, the highly successful journalist who was a correspondent of *Corriere della Sera.* [36] In July 1962 Montanelli contributed five articles fulminating against ENI and Mattei. According to Montanelli the argument about ENI had in the course of time developed into a theological dispute in which participants were divided into the two sides of the 'faithful' and the 'unfaithful' to the master, with nobody being able to adopt an attitude of indifference. Montanelli first emphasised the political aspect of the enterprise, denouncing the paradox that this company set up to serve the State had in ten years achieved mastery over the State, 'holding the keys to a large Italian treasure chest' and negotiating with the Russians without any diplomatic assistance whatsoever. (This could be regarded instead as evidence of the dynamism of the enterprise and as a kind of recognition by vice of virtue.)

Montanelli then claimed that political support for Mattei came from the fact that Vanoni, the Minister responsible for the National Plan, was a convert to the Christian Democrat cause from socialism; nor was the well-informed Sr de Gasperi, as Chairman of the Regional Council, inclined to disfavour Mattei's activities. The idea that the Americans might obtain a firm hold in the Po Valley industrial area would inevitably produce nationalistic over-reaction in the population, especially in its Communist part. It was not a bad thing to give them instead a former 'partisan' acting in the name of the State and of anti-capitalism, who was therefore scarcely open to suspicion of being a 'colonialist' and 'imperialist' or an 'exploiter of the people'.

Montanelli's judgment was less shrewd when he condemned the barter agreements with the USSR. He claimed that the economic cost of the transaction could not be assessed exactly, because the products supplied in return were such items as foam rubber and nobody knew the correct price. This highly polemical argument which exaggerated the difficulties of calculation, was purely tendentious, omitting to mention the positive effects, such as improved bargaining power and great increase in investment, deriving from a deal of this kind. The attack on ENI policy abroad was also rather weak. It was easy to be sarcastic about Mattei as 'an oilman without oil' because Mattei's efforts in Iran, in Egypt and in Morocco tended to justify that idea. The journalist's line was that ENI ought to trade in oil not produce it, but the enterprise was trying to use trade as a stepping stone to becoming a producer. (This method may be irrational from the point of view of cost but it is at least logical in an integrated oil company as had been intended by the Act of 1953.)

Again, in the matter of price there was complete disagreement between the correspondent of the Milan newspaper and the head of ENI. Montanelli expressed the matter pointedly, saying that ENI 'lost a milliard on petrol and made four milliard from methane'. Montanelli asserted that as a result of the monopoly in the Po Valley, ENI made a net profit of 7 lire per cubic metre of methane, which represented more than 50 per cent of a selling price of 12 lire. He thought, contrary to the deliberate intention of the Act of 1953, that the price of methane was too high. (Vanoni had explained at the time that the reason for the high price was not to give a very unfair advantage to the North, which was already favoured by nature and by history, to the detriment of the South, which had always been unfortunate.) In his reply to these insinuations, [37] Mattei disputed the figures given by his antagonist and said that the price of Italian methane was 'one of the lowest in the world', mentioning that this was 'a fact that can easily be checked'. (The truth seems to be halfway between these two claims. The price of methane is calculated by reference to the price of oil; it is therefore fairly high and undoubtedly gives a good profit margin. But this was the important bonus which enabled ENI to grow and as a result of growing to achieve a stronger competitive position so that it could exert enough influence to bring oil prices down.)

Montanelli at least had the merit of posing sharply the question of the overall cost of this hydrocarbons policy. He claimed that it was impossible to assess exactly how much the nation had to pay to keep its infant prodigy oil enterprise, because ENI gave little guidance to the observer when it issued group accounts. It was difficult to tell from the accounts published what was set aside for reserves, what was investment and what was provision for repayments. Mattei's reply, in the same newspaper, [38] adopted the legalistic stand that the Act of 1953 had required a balance sheet for the various activities of ENI to be drawn up, that the general legal rules had been specified and that the consolidated accounts were investigated in the normal way by the public authorities. ENI obeyed the rules laid down. He added, moreover, that on each occasion Parliament considered the balance sheet for ENI and the explanations of the Minister. Parliament could exercise rights of control by means of motions, questions and formal requests for statements. Mattei said that Parliament had made a great deal of use of its rights, because 'some 500 questions, formal requests for statements and motions have been put down'. Montanelli countered that the reply was very formal and quite a long way from the truth of the matter. [39] He was not far wrong in this, because in all democracies the control actually exercised by Parliament is somewhat theoretical and ill-adapted to the variety and complexity of situations occurring in business life in the modern State. Montanelli's point was corroborated a few weeks later [40] by material from a Deputy in Parliament, who produced a copy of the text of two of his questions put down in September and November 1961 for answer in the Chamber of Deputies. The Deputy commented that these were 'doubtless embarrassing but highly relevant' and that 'no reply to them has been given, despite repeated applications and notwithstanding the very precise rules of procedure of the Chamber'.

The excerpts quoted above give some idea of the climate of mingled suspicion and enthusiasm in which ENI has grown up as a state company.

Mattei showed such a forceful personality that he could not fail to provoke strong reaction in the over-charged atmosphere of the post-war period. The opposition encountered indubitably forced ENI to state its objectives to public opinion and put more of its cards on the table than it wished to show. Since its founder left the scene and in an international climate that has been transformed by the end of the Cold War, ENI has lost its initial originality of character and is less controversial. But its experiences during that period have given it the habit of a semi-clandestine way of life.

The degree of secrecy surrounding the French ELF group is perhaps even greater. It was only after the crisis of July 1970 with Algeria and the virulent attacks of *El Moudjahid* that the French enterprise decided to publish some precise figures in connection with the cost of its operations in Algeria, the sums paid to the Algerian Treasury, the profits per ton produced, etc. A fact which emerged was that in compliance with the 1965 agreements ELF/ERAP handed over in Algeria an exceptionally large proportion (85 per cent) of its gross receipts, which was a much higher proportion than applied in any other case in the producer countries at that time. But throughout the Franco-Algerian negotiations of 1970–71 secrecy was rigidly observed. This seems regrettable because of the fact that a public enterprise was involved. Moreover, it is always useful in difficult negotiations to be supported by the weight of public opinion already pleading the case as strongly as possible. Instead there had been scant understanding of ELF's policy, even in the very branches of the administrative authorities that might have been expected to know what was happening, there had been failure to appreciate the efforts made, and there had been ignorance on the part of public opinion, which had little inkling of the facts of the matter or of how hamstrung the company had been. Granted, there is a tendency in France to come to blows on matters of principle and a great deal of controversy arises about the idea of public enterprise. But apparently with ELF, as with other state oil companies, there has been a great change from the traditional way of acting.

Danger of ambiguity in the status of a public enterprise

The difficulty of exercising genuinely democratic control over public enterprises creates some doubt about certain established dogmas of socialism. Professor J.K. Galbraith has pointed out that in the 1960s many socialists concurred with the view of a senior figure in the British Labour Party, C.A.R. Crosland, who wrote in 1965 that public corporations were, by nature, 'remote, irresponsible bodies, immune from public scrutiny or democratic control'. Professor Galbraith said that 'socialism has come to mean government by socialists who have learned that socialism, as anciently understood, is impractical'.[41] Thus, public enterprises are accused of not being what they should be and of possessing a status incompatible with their actual practice.

But does this contention, in fact, hold good as regards the oil business? Is a state company the first stage towards socialism throughout the economic apparatus, an experiment in democracy, a bridgehead for protecting the interests of the nation or a means of assuring more rational conduct of business in the existing economic system?

Because of its importance oil cannot remain outside politics. On the other hand it is subject to such strong competition that its extraction, processing and sale can be performed only by companies which are prepared to enter vigorously into the industrial and commercial worlds. Therefore a state oil company cannot help facing two ways.

Algeria envisages achieving socialism through Sonatrach but in reality that company would fit into the system in a capitalist state fairly well, since it seeks to promote patriotic endeavour and aims at making profits in the interests of its members and for the benefit of the community. Despite the socialist ideology claimed for it, Noc seems to be following the same path.

Since the last world war Italy has invented bold formulas for combining private and public interests and involving the State in economic development with the minimum of rigidity. ENI is just one star of first magnitude in a constellation containing several others and its behaviour shows no fundamental difference from that of the other enterprises, whether they are public or private. Similarly, in Spain there is no great difference in the way in which Hispanoil functions compared with other enterprises which are partially state-owned or which have been formed with a majority holding by banks.

In France, where debate of the principle of the state enterprise remains lively, uncertainty about where a public enterprise fits into the scheme of affairs is perhaps even greater. An official report, the so-called Nora report, helps to show what the situation is.

ERAP is a curious case among French public entities, so much so that the Nora report is unsure how it fits into any scheme for identifying their status. The report expresses clearly the difficulties about the status of public organisations:

> The composition of the public enterprise sector, in which outlines are blurred, does not seem to conform to any single principle of

economic or political logic . . . The confusion regarding status prevailing in the public sector derives both from disparity of features and from divorce from reality: there are management services, public administrative offices, public undertakings, national enterprises, state companies together with their privately-owned affiliates, and partially state-owned companies.[42]

Despite the difficulties, the report then tries to apply some classification to public enterprises and puts ERAP into a fairly precise class:

> With the notable exception of the 'national companies', which are nationalised loan and insurance enterprises whose corporate capital is wholly publicly-owned but whose management is on identical lines to that of private enterprises in the same sectors, a fairly clear distinction can be seen between enterprises whose status is that of a 'public undertaking' and those which have the form of a partially state-owned company. But assignment of public enterprises to one or other of these two systems creates inconsistencies. There is no correlation between the scale of the public service function in the operations of each of these enterprises and their inclusion in one category rather than the other: the energy enterprises (EDF, GDF, Charbonnages, ERAP) are all public undertakings but the transport enterprises (Air France, SNCF) are as a general rule partially state-owned companies (although RATP is an exception).[43]

The report goes on to say that if the criterion chosen to characterise nationalised economic operations is that of heavy industries distinguished by great concentration of production units, that definition is meaningful as regards electricity and coal but is inapplicable to oil affairs and the iron and steel industry. Division into monopolistic sectors and sectors subject to competition is no more helpful: the nationalised concern EDF is indeed a monopoly but ERAP and Renault are also nationalised, yet they are both at the centre of the competitive fray'.

The report's distinctions are of no help in determining a specific status or mode of functioning of the French public oil sector as a whole, since that sector is comprised of a state enterprise, ERAP, and a partially state-owned company, CFP. The reason for the form assumed by CFP is that an international consortium was set up, therefore the state could not be the sole shareholder of the company. The date of incorporation (1924), the way in which the company was to function internationally and the absence of clear conceptions of state economic services rendered a wholly nationalised oil undertaking unthinkable and impracticable. In any case, the tasks set for CFP from the outset put it fairly definitely into the class of partially state-owned enterprise. The position as regards the status of the other oil company is quite different. In any classification of the wide and varied range of French public entities ERAP is bound to occupy a place apart, defying assignment to any precisely defined category. It is state-owned but sound management is no less necessary than for private enterprises in the same line of business. Operating in a key sector where a constant flow of heavy investment is required, ERAP has no monopoly and is subject to competition. This is precisely the aspect of its status that seems most interesting and it is a pity that the Nora report did not examine the oil sector in greater depth. After all, ERAP is a modern type of public enterprise satisfying a principle regarded as being of fundamental importance by the commission which produced the report, namely that nationalised sectors must adjust themselves to the requirements of competition:

> International competition creates an obligation for the French economic system, that of competitivity. It is not only a matter of abundant production but also one of minimum price, or more precisely the latter is a condition for the former . . . True, this obligation weighs more heavily on industries exposed to outside competition than on those which are protected by the nature of their activities, hence in the nationalised parts of the economy on coal more than on electricity production or domestic transport services. But in the final analysis the progress of the protected industries directly affects that of the exposed activities. Thus we are no longer in the era when the essential task of the nationalised sector was at all costs to perform the operations for which it was set up. Now the task is achieve greater productivity.[44]

Certainly coal is dramatically exposed to international competition but this does not serve as a good example of what is happening at present in nationalised energy production. A branch of the economy that is in recession, as coal is, does not serve as a satisfactory guide to the conditions of viability of an 'exposed' public enterprise. The oil industry seems to be more relevant to the issue. This is an expanding branch of activity with progressive annual increase in its share of provision of the total volume of energy required to meet consumer demand. An oil concern is under all the greater pressure to improve its productivity for operating in a competitive situation in a field in which demand is constantly rising. Oil might even be said to be doubly 'exposed' to competition. Firstly, like its former rival coal, it is exposed to competition between the various energy sources for ability to provide the most economic supply. The price must be competitive but at the same time it must repay the heavy investment necessary. The ELF/Aquitaine group is the second largest investor in French production, the largest being EDF. Secondly, whereas there is a monopoly in coal, there is none in the French market for oil and therefore there is the same struggle for survival and for success in order to advance further as under a capitalist system.

This effort to increase productivity is producing a new kind of public enterprise that did not exist at the time of the nationalisations which followed the Second World War. The predominant consideration then was production at all costs. If profitability proved to be unattainable, then communal authorities made up the deficits of public enterprises, since the latter were regarded as providing a public service. The Nora report reflects a change of attitude. Henceforth the emphasis was to be on profitability and on exposure to the hardships of both domestic and foreign competition. This competition is perforce introduced by greater French participation in international trade. However, suddenly circumstances are such that the whole 'socialist' conception of public enterprise is challenged.

It seems obvious nowadays that even a public enterprise should aim at profitability and not rely on aid from communal funds to make up the deficits due to mismanagement and unwillingness to take part in a struggle with competitors. Yet it is normal for the state to subsidise state companies in proportion to the burdens which it places on them by requiring them to fulfil purposes which it regards as being in the national interest. Therefore there should be greater discrimination between the part of the price of items supplied which derives from economic rationality as commonly practised in the relevant branch of business and that part arising from the measurable cost of political decisions.

Thus, in France's co-operation with Algeria in oil affairs it would have been advisable to differentiate more clearly between the investments incurred by ELF as an oil company operating in a competitive situation and what was attributable to the French Government's political policy of special co-operation with the Algerians. In order to avoid confusion, ERAP's financial effort should not have been the result of the sum of those two requirements. The company could not simultaneously be an economically profitable enterprise and the means of state generosity to Algeria. The State ought to have ascertained how much aid, including that in connection with oil, it proposed to provide for Algeria and such aid should have been supplied through government channels. In those circumstances the French company would have made its investments in accordance to its overall policy, without impairment of its competitive powers by tasks which were disproportionate to the nature of the company. In any case, if for political reasons co-operation with Algeria was to be maintained or increased, the correct method was not to subsidise ERAP massively in order to make it a special vehicle for that endeavour. This was going back to the 1945 way of handling situations that was now supposedly to be renounced and it meant treating ERAP in the same way as the public transport services or the railway system. The procedure was contrary to the concepts of the work of state companies which had already evolved; it ignored the promotion of industrialisation under the Sixth Plan and it disregarded the intentions of the Nora report. ERAP's ten year adventure in oil in Algeria has been described by the head of the group, Pierre Guillaumat, as a 'break-even operation'; there was simply recovery of the capital, 'without interest and without increase in value, or only a very small addition'.

Danger of indiscipline in financial affairs

The state companies under special consideration in this book entered the international scene late and found it already occupied by well organised international groups which were by then firmly entrenched in many regions of the world. Adopting a policy of integrated operations, they were obliged to pursue a particularly bold investment policy if they were not to be overwhelmed by the competition.

At that time the majors had very well balanced financial arrangements, because they were capable of self-financing to a large extent. The state companies were unable to match that situation. In the 1969 financial year ELF/ERAP made investments totalling 624·4 million francs, compared with 567·9 in 1968 and 644·4 in 1967. For this purpose the company drew on funds of 704·8 million francs in 1969, 680·9 in 1968 and 659·3 in 1967. Its own investment funds in 1969 were 349·6 million francs, comprising 261·3 for self-financing, the remainder coming from repayment of loans to its affiliates (51·2) and from transfer of assets or other capital resources (37·1). Funds from outside sources totalled 355·2 million francs, comprised of loans of 50·2 million francs (medium-term loans of 71·7 less loan repayments of 21·5), a subsidy for underwater exploration and sundry operations of 17 million francs and a grant of 288 million francs from the public hydrocarbons fund.[45] It should be noted that two-thirds of the loans (49·6 million francs) for financing investments abroad were contracted in foreign currency, partly because of a shortage of available funds on the French money market. In 1970 the overall turnover before taxation of the ELF group, including SNPA, was 7·7 thousand million francs, compared with 6·8 in 1969. Investments rose from 2·2 thousand million francs to 2·7. In 1971, following events in Algeria, the total dropped to 2·1 thousand million but in 1972 it recovered to 2·6 thousand million. It should be noted that the turnover before taxation of SNPA and its associated production companies abroad (totalling 1·29 thousand million francs, compared with 1·18 in 1969) does not include the excellent turnover before taxation of Aquitaine-Organico, at some 436 million francs (compared with 378 million in 1969). ANTAR had a turnover of 1·5 thousand million francs for 1970.

Figures for the funds being applied by ERAP show that self-funding is steadily increasing (169·8 million francs in 1967, 227·1 in 1968 and 261·3 in 1969) and that the proportion of self-funding in total investments improved in 1969 compared with 1968 and even more so compared with 1967. But at the same time sums borrowed became larger, except in 1968 (33·4 million francs in 1967 and 71·7 million in 1969 but only 9·9 million in 1968). For the ELF group as a whole the gap after self-funding was 1·4 and 1·5 thousand million francs in 1971 and 1972 respectively. But in 1972 it represented 50 per cent of the funds required, compared with 55 per cent in 1971.

Meanwhile there was a steady reduction in the grants made by the public hydrocarbons fund:

State grant in 1967 . . . 363 million francs
" " 1968 . . . 350 " "
" " 1969 . . . 288 " "
" " 1970 . . . 250 " "
" " 1971 . . . 200 " "

Thus, the subsidy has declined by 55 per cent over five years and in the grants since the company's difficulties in Algeria the Government has not chosen to increase the sum. This strict treatment shows that the Government wishes ELF to become self-sufficient as quickly as possible and dispense with the crutches which it still needs at present. However, it is difficult to see when the company will achieve that position. There has been mention of 1980 as a reasonable date. The speeding up of Government reduction of its subsidy suggests that an earlier date is envisaged. What happens will depend on how well the group succeeds during the next few years in making up its losses in Algeria and how quickly it is able to find fresh sources of crude in the world and thus offset its heavy expenditure on exploration. In the meantime it is difficult to see how the state company could manage without that subsidy, which is comparable to an increase in capital in the case of a private enterprise, because in 1968, for example, it was equivalent to more than half the total investments made by ELF/ERAP.

The French Government's strictness towards ELF contrasts with the attitude of the Italian Government, which has been keeping up a high volume of aid to ENI. In 1968 it undertook to make no reduction in a loan of 250 thousand million lire over five years at 50 thousand million a year (575 million francs). Converted into francs this makes the 1972 subsidy nearly twice the amount granted by the French authorities in the same year.

It must be acknowledged that the government subsidy is of great assistance to ENI, which is rather heavily in debt as a result of its ambitious world-wide expansion efforts. In 1959 only 37 per cent of its investment funds came from outside sources but in 1962 short-term debts had risen from 19·1 per cent to 27·1 per cent of the total debts. Given this financial position of the company, British economists were able to comment in 1967 that the source of ENI's troubles was not so much the mistakes or excessive ambition of its earlier days but the fact that 'it has set out to pursue entrepreneurial policies with a capital structure better suited to a municipal gas company'.[46]

The situation has improved somewhat since that date but it is still much worse than that of its French counterpart. For example, the 1969 balance sheet showed medium- and long-term debts amounting to nearly a third of the total liabilities and 12·4 per cent higher than in 1968. The largest item in the liabilities was loans, amounting to 683 thousand million lire in 1969 out of a total of 2,706·3 thousand million. Interest therefore absorbs a large slice of the group's income. The Chase Manhattan Bank has made an analysis of the use of funds by ENI and by some United States oil companies. The bank has found that with an index of 100 for the total receipts the proportion of working costs was about the same. But the level of interest payments was 1 for the American group and 8 for ENI. Similarly, repayments were 8 for the former and 15 for the latter.[47]

Thus, the financial structure of ENI differs markedly from the mean for international groups. However, the investment burden is changing. Even the majors are being obliged to contract heavier debts. If the trend continues, then the financial structure of ENI will look less aberrant.

The accounts published by ENI contain several surprises, including the smallness of the profits in proportion to the size of commitments and the high ratio of foreign funds to domestic funds for financing investment. It looks as if ENI may have become too accustomed to a continuing flow of government aid.

Aid is, of course, necessary to enable a new state company to become established but it seems unsatisfactory that the Italian Treasury continues to receive a very low rate of interest on the capital advanced to the state company. The rate was 0·03 per cent in 1967 and 0·3 per cent in 1968.[48]

Similar findings might be obtained for Sonatrach but investigation is easier in the cases of ELF and ENI, for which accounts documents that scarcely exist in relation to the Algerian company are available. Even with the two European companies a great deal of truth about the oil business escapes scrutiny. The view of their operations obtained in their own country is not clear and their international activities are even more obscure.

Danger of political impediment at international level

Agreements between state companies either side of the Mediterranean have been made under pressure from the respective governments. The intention was to demonstrate that arrangements between producers and consumers could be made more directly than through the traditional intermediary of the large private groups. The independence and the respective interests of both sides were thought to be better assured by that means.

Oil seemed to be a vital commodity so exposed to political manoeuvre that if it were left entirely in the hands of private enterprise insoluble problems might arise in future. Italian and French officials reached similar conclusions on this point. As head of ERAP, Pierre Guillaumat once explained that the security of France's oil supplies could be threatened by three situations, namely when French general policy displeased the foreign oil groups, when the policy of the latter – which is decided independently of France – displeased the Arab producer countries, and when French policy displeased the producer countries. The first two dangers were thought to be averted by the possession of national facilities for protecting oil supplies. ERAP and ENI have sought to overcome the third threat by means of contracts or deals with the producer countries that were particularly advantageous to the latter, so that they would never be tempted to depart from the agreed arrangements. The Governments on either side have endorsed this system and it is hoped that the former competition between the three sides participating in oil affairs can be modified to the advantage of the state companies. Possibly the project has been performed too quickly, taking a favourable opportunity to promote what is rightly a long-term achievement and risking a strategic move when world strategy was relatively immobile.

The agreements made by Hispanoil, ENI, Aquitaine and ERAP in Libya have been designed for this purpose. They seem to be relatively unkind to buyers of crude but their effect cannot be properly judged until the tonnages under these new arrangements are substantial enough to be noticeable in the world market. The Franco-Algerian pacts of 1965 were of a different nature, constituting a step in a new direction. For the first time oil affairs between a consumer and a producer country were explicitly dealt with at political level under an agreement demanding a new style of co-operation between the two state companies.

It must be admitted that the experiment was somewhat inconclusive, because it has not been fully tested. In return for credits and loans France was able to obtain supplies at a particularly advantageous price, especially when market prices began to rise sharply. But three years after its commencement the agreement between the two parties was already at issue and the terms continued to become higher until the 1970–71 negotiations and Algerian nationalisation of oil in spring 1971. The security of supplies which France had hoped to ensure proved to be as uncertain as ever. ERAP and the French Government versus Sonatrach or the Algerian Government were no better placed than an American or British private company dealing with a recalcitrant government.

It is difficult to put the problems into purely economic terms nowadays. There is a danger of politics invading the whole of the economic field.

Where there is an impasse because of the contradictory nature of the interests involved, political intervention might normally be a means of deciding the matter. But if negotiations have already reached political level there is no new or higher facility for making room for compromise. There is a paralysis which can only be overcome by a complete revision of relations between the two countries. Such a situation cannot occur when private companies negotiate on their own behalf.

Would it therefore be better to restore an intermediary function to the private companies? The Chairman of the French company in the BP group, J. Huré, must have been thinking along these lines when he said in 1966:

> Between producer countries and consumer countries, whose interests are divergent (the former seeking the highest and the latter the lowest prices possible), it is advisable, as in the case of areas of friction, to include a lubricant in the form of private companies outside political struggles. They belong to both categories of country and it is to their advantage that each side is satisfied, therefore the groups are well placed to find compromises between the respective interests whereby crises can be prevented.[49]

It is strange to hear that private companies are outside politics. They are involved in politics whether they like it or not, and it is best to acknowledge that fact. But the important point is whether or not it is advisable that countries of the two camps separate economic affairs from a system in which the State acts as final arbiter. As P.H. Frankel said in 1948, 'if decisions of a strategic nature can only be decided at the level of the supreme authority, then it is important that tactical decisions are left to the industry itself'.[50] The approach adopted by the British has tended to differ from that chosen by the French; encountering growing hostility to their national interests in all the Arab countries they have preferred to allow BP, which is nevertheless state-owned, to place strong emphasis on its commercial nature. For France, CFP has taken up that line and does not seem to regret it. The obstacle of politics impinging on economic affairs has been greater in the case of ERAP. The political factor may have been helpful when the market was fairly competitive, favouring buyers. But when circumstances changed and favoured sellers the inflexibility of the arrangements made with the intervention of politics became a disadvantage and individual decision-making was swept aside. Algeria invoked three means of winning what it wanted – OPEC, the country's geographical position in the Mediterranean and the special relations with France. The Algerians wished to obtain both a rise in the posted prices and taxation at the level which had been reached in the Persian Gulf states, they wanted special advantages because of the proximity of the country to Western Europe, at the same time they aspired to 'naturalisation' as far as possible of the crude extracted by French companies, whether these were state or private companies. The French wanted to continue to negotiate with Algiers but they also sought to safeguard their interests in the Sahara and not to be too far out of line with the terms which the large companies and the independents would be obliged to grant to producer countries belonging to OPEC.[51]

The producer countries have used collective agreements to try to obtain heavier taxation, increased prices and a larger share of operations in producing and handling national oil. They have the wider long-term aim of completely different sharing of the profits from oil operations.

As figure 18 shows, in 1969 the proportion of tax levied by the European consumer countries was six times the proportion received by the producer countries. But this ratio was only valid in the case of petrol, which was heavily taxed (especially in France and in Italy), and not for fuel oil, for which less severe indirect taxation took into account its use as energy for industry. Fuel oil, is of course, used in much greater quantity than petrol in the European countries. In fact, petrol amounts to only

15–20 per cent of the combined consumption of the two items.

Thus, in its 1969 report on the French oil industry, DICA, the Hydrocarbons Division of the French Ministry of Industry, treated taxation as a separate item affecting the use of motor vehicles more than the oil industry. According to that report the mean selling price after taxation in Europe was made up as follows:

– direct production costs: 5 per cent;
– taxation in the producer country: 10–15 per cent;
– transport costs: 5–10 per cent;
– refining costs: 10 per cent;
– costs of supplying to the consumer: 50–60 per cent;
– cash-flow of the oil sector: 10 per cent.

The figures given by DICA and by the French oil trade association, reveal approximately the same percentages, although DICA omits to mention the rate of taxation in the consumer countries. The breakdown of the price makes the share obtained by the producer countries at that time look small but it might be argued that crude oil has no value in that form: it has only become a sought-after commodity because Western

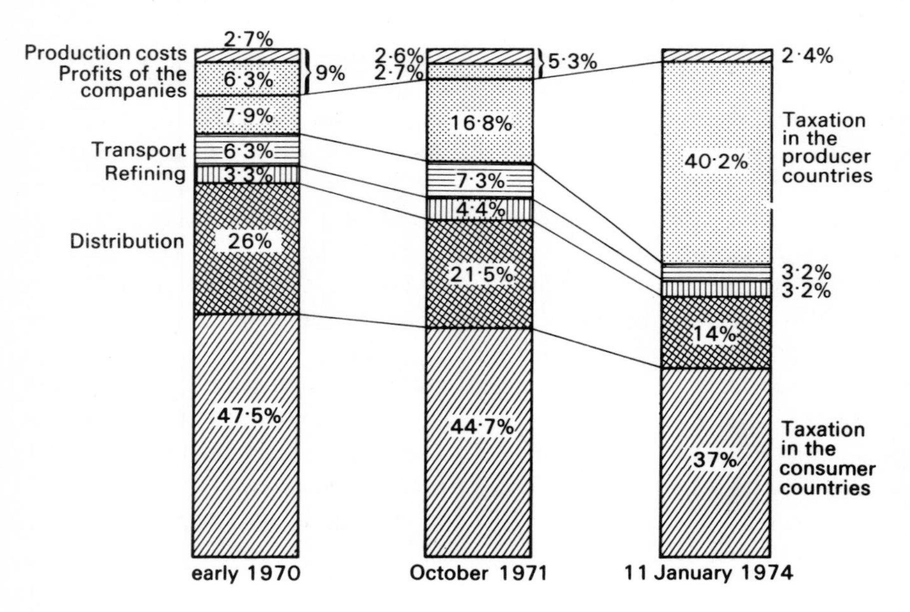

Fig. 18 Breakdown of funds involved in oil operations

Source: Comité professionnel du pétrole, *Pétrole 1973*, p. 5

technology was able to process it and give it a large amount of added value.

Processing is precisely the stage of operations in which the producer countries wish to intervene. Dispute between producer countries and consumer countries largely relates to the sequence between production from the well and consumption from the pump. The importing countries and the enterprises supplying them still have control of the stages between those two ends of the series of operations. The producer countries would like to take over that sequence of downstream operations, but redistribution of those important activities will require time, technical skill and capital and it must inevitably reduce the involvement of the big companies in oil affairs. However, as companies perform the greater part of these downstream operations in countries of the Western Hemisphere, it is difficult to see how the state companies of the consumer countries could fail to support the big companies in seeking to preserve that situation. The big companies still form an unavoidable bridge between the producers and the consumers of the world, even if their influence is gradually being whittled away. The producer countries achieved a great victory in Teheran and the example of Sonatrach, which is the only producing company in an Arab country to obtain control of output capacity of more than 30 million metric tons, may eventually evoke imitation. But the producer countries have not yet made much progress towards their final objective. The host countries can, of course, take over the crude oil in their subsoil, if they have sufficient finance and technical knowledge to extract it, they can nationalise foreign interests operating in their territory, but they are not yet capable of successfully performing the subsequent stages of world oil business.

In the circumstances, their temporary objectives may resemble those of the big companies, in that a sufficiently profitable price to enable exploration to be worthwhile is required. The fact that these participants in affairs are in accord in that respect may prove disadvantageous to European consumers. The only way in which the state companies of the European countries can counteract the situation is to introduce carefully planned reforms on a basis of gradual co-operation, in the hope that circumstances may change in their favour.

In this section, five dangers most likely to arise in connection with state companies and to be produced by their ambiguous nature have been indicated. The potential or actual faults of these enterprises do not necessarily mean that their existence should be condemned. The real issue is whether, despite the obvious or concealed defects which may be revealed by critical examination, state companies can play a useful part in world oil affairs. Are they now or in future capable of improving on their past performance?

Raison d'être of state oil companies

Because of the deficiencies which the state oil companies tend to possess in greater or lesser degree, people from time to time question whether these companies are necessary to the success of realistic hydrocarbons policy and whether it would not be better to dispense with their unsatisfactory conduct of matters which other forms of undertaking manage better. To decide this point it is necessary to investigate the usual reasons for setting up public enterprises and their value in protecting consumers, before proposing conditions for the survival of state oil companies.

Usual reasons for setting up state companies

A fundamental question arises in connection with state companies, namely what is the *raison d'être* of public enterprise in non-communist countries and outside the public utility field? Does the government, to make its weight felt, have to enter the ring and become a combatant rather than remain the referee? Why can the government not be content with general measures to oblige entrepreneurs to obey certain rules of the game? [52] In the European countries with Mediterranean coastlines one finds that the government has begun to take specific action by protecting or by penalising individual enterprises participating in oil affairs, thus becoming judge in its own case. Not content with creating a general system of regulation and confining itself to certain national needs, the government has turned itself into a competitor of its subjects by performing managerial functions and by assuming charge of the financing of new enterprises. Why has this been done? There are several possible explanations, ranging from reaction to a position of dependence to an ethical intention and including demands of economic rationality.

Earlier in this book it has been pointed out that experience in wartime or in time of crisis may produce awareness of a situation which was of less significance in peacetime. The setting up of a state enterprise in such cases may be regarded as a national reaction to the undue influence held by some large companies or producer countries occupying a position of politico-economic power, as a result of which they may be able to threaten the sovereignty of a country. The danger that a régime or a few specific groups could hold a near-monopoly or a dominant position in a key sector such as energy must be averted. In such circumstances, if less powerful operators do not take suitable precautions they may be forced into merely subsidiary operations yielding little profit and putting them still further below the group of the strongest participants. The cost in foreign exchange of oil supplies for the countries of Europe should also be mentioned: lacking natural deposits in their own territory, their demand becomes very heavy as industrialisation progresses and as the proportion of energy supplied by the domestic coal resources declines. If there is a failure to react to such a situation early enough a nation's debts pile up very quickly and if the trend is not brought under control there will be increased dependence on a few countries whose currency rules international markets.

Such basic motivation, even if there were historical events which were also relevant, was the main reason for the formation of CFP, ERAP, ENI, Campsa and Hispanoil. With remarkable similarity the pattern was repeated in Algeria, Libya, Morocco and Tunisia, in a delayed reflection of European attitudes. Surely Sonatrach and Noc were set up in order to try to eliminate the 'backwardness' of which an official statement complained, to react against systematic pressure from dominant forces in the Western Hemisphere and to enter in turn, with their own means, into the commercial and industrial adventure of oil affairs. But history has demonstrated that the length of time for which interested parties have been actively engaged in oil operations exercises a decisive influence on the success achieved. Oil policy requires a long period of planning. Investments in oil operations are extremely slow to produce definite results. The later a participant starts, the steeper is the climb to reach the current level of affairs and the more uncertain are the chances of success. There is rivalry between the individual participants but they do not all start off with equal chances and those who are in the lead naturally wish to prevent newcomers from catching up.

There is a further consideration, which arises directly from conditions in the country's economy. One reason for setting up a public enterprise has often been the non-existence of private enterprise in the field of oil operations or its failure to assume the risks involved. The function of public enterprise in that case is to fill the gap. This is the case with the countries of North Africa, where in addition to the rôle of supplying the required facilities, the state oil companies have now been given the task of replacing foreign participation in activities. The Italian AGIP, the Spanish Campsa and the predecessors of ERAP, the French undertakings BRP and RAP, were likewise created to meet a need. At an earlier stage in the history of oil business, the main justification claimed for bringing state enterprises into being was to apply an anti-trust principle to the market and promote a healthy state of competition. Dr P.H. Frankel has argued forcefully that the introduction of a public oil sector in Europe could act as an equivalent of American policies of applying measures to prevent the opportunity for abuses provided by domination of a market by a few large operators. He has quoted a comment that public enterprise, especially in the oil sector was 'a European version of the American anti-trust philosophy'. Its purpose was 'to provide countervailing elements against agglomeration of economic and political power in the hands of large-scale private enterprise'. He has noted that, according to an observer in the United States, 'The Supreme Court has expressly recognised that the fragmented economy is desirable . . . even though inefficiency may result'.[53]

Public enterprises make much play with this argument, claiming that the government's rôle is to establish a situation in which the various operators can 'compete with each other on equitable terms'. This leads to the further proposition that state financial aid must continue and increase until the imbalance between the public and private spheres of the economy has been to a great extent eliminated.

Should the conditions for fair competition be achieved when state organisations are present, then the advocates of public enterprise claim that it must remain in being to serve as a corrective agency, acting as a means of stimulating endeavour by private enterprise, of obliging the latter to obey certain rules of the game and of ousting it from exceptionally favoured positions. But to succeed in that situation it is necessary, given the competitive climate, for public enterprise to have a healthy financial position and to be better managed than private enterprises. It is rare indeed for public enterprise to achieve those standards.

In practice, therefore, the arguments advanced in favour of state companies are rarely mentioned in relation to oil affairs, in which the distinguishing features of public enterprise are likely to be blurred by the necessity for sustained competition. In such circumstances it is preferable to give a state enterprise a certain degree of ethical tone. In France, but not to any great extent in Italy and still less in Spain, the managerial functions in state companies are performed by personnel who have come from public administration or from the leading technical colleges and not from private enterprise. In Italy and Spain there is a tendency for all business concerns to look very much alike, being so organised that there is little scope for distinctions between private business ventures and those enterprises in which the State is involved. Moreover, there is not such a strong public service tradition in those countries as in France. In the countries of North

Africa the industrial and commercial apparatus does not yet provide sufficient material for the comparison of the two systems: both the public and the private spheres of modern business affairs have yet to be created but in the recruitment of managerial personnel there are signs of a tendency to copy the French model.

In the higher echelons of the French state companies there is a conspicuous preponderance of administrators inspired by the mystique of the public service tradition, seeking to work for the interests of the nation as a whole, anxious to put commercial profit into the coffers of the nation and not into the pockets of private capitalists. This is an attitude frequently found in France, at any rate in the generation now at the helm. The phenomenon may have something to do with the reservations about money held by some sections of the Christian faith. A similar outlook is widespread in Italy. According to Giusseppe Petrilli, head of the Instituto per la Ricostruzione Industriale (IRI), public enterprise stands at a vital intersection in the economic world, because without sacrificing the conditions required for efficiency it must fulfil more general needs described as 'political'. Petrilli has stated that: 'The ethics of public enterprise could be said to match those of planning, if that term is taken to mean a type of intervention in economic affairs which seeks to influence and orientate the quantitative and qualitative development of the economy by means of moves designed to shift emphasis in the market, and hence of decision-makers in the sphere of economic activities, without actually taking direct action to change the system of ownership and the functioning of the market but instead applying the principle that the functioning of the market is an infallible guide to the efficiency of the management of enterprises and to the productivity of the public and private investment involved.'[54]

Petrilli's description of the complex purposes of public enterprise seems to be valid for state oil companies. The best justification for their existence is provided when at the same time as working efficiently they avoid adopting an attitude of expediency – of merely trying to suit their operations to 'the interests of the market'. Their activities should be closely guided by market conditions, but in the sense of paying attention to the market for the purpose of exerting on it the influence of the interests of the nation as a whole – that same burden of responsibility which sometimes impedes their functioning.

Protection of consumers

Is there not a risk that if state companies are fully-committed participants in business competition this will gradually cause them to lose sight of a purpose which was a very important reason for their existence, namely that of protection of consumers?

The top men of public enterprises of the countries of North Africa proclaim that they are very anxious to pay due regard to the interests of consumers. According to what they say it would seem that the big companies are nothing but an unnecessary barrier between buyers and sellers and that an alliance between state companies with headquarters either side of the Mediterranean would give results entirely satisfactory to both sides. Everything suggests the contrary. The producers tend to want to obtain the maximum income from extraction and processing of oil. The interests of the final users bear little weight against the natural inclination of both supplier and purchaser states to acquire ever-increasing revenue from petroleum products. Can the same contradiction also be found, despite protestations to the contrary in the case of state companies in consumer countries? Usually, the more successful a public enterprise is, the more it strives to achieve integration. So long as it represents the troubles of consumers at the end of the chain, that state company desires the lowest possible price for energy. But as soon as the same company becomes a producer, its ambition is, of course, to obtain a high price, sufficiently remunerative to give the company a firm basis for expansion. One therefore wonders whether or not it turns out that consumers always necessarily suffer if a state company achieves economic success and becomes more dynamic.

In a thought-provoking article entitled 'The current state of world oil',[55] the British oil expert Dr P.H. Frankel has shown some scepticism about the effort undertaken by some large consumer countries to obtain crude resources of their own. He asserted that 'an attitude of that kind, adopted first in Italy, then in France and in Japan, either by state enterprises or with government backing, far from achieving its aims, would seem to have done a disservice to the national interests which it was intended to serve'.[56]

In support of what I feel obliged to call a thesis, the British expert painted a picture of a situation very different from the state of affairs earlier in the history of oil. At the time when the majors were receiving a very large profit margin, of the order of 70 cents per barrel, the other oil companies and the governments of the consumer countries would have liked direct access to the sources of crude 'at cost price'. This would have obviated a 'toll' or 'due' destined to enrich certain profiteers and would at the same time have given the consumer countries rights to very valuable commodities.

Dr Frankel gave four main reasons why events have not been as favourable to consumers as that arrangement would have been:

Competition had caused a two-thirds cut in the profit margins of the oil companies since that time. The 'super-profits' which continued to exist even after the adoption of the fifty-fifty profit-sharing scheme had now almost disappeared, thus the main reason for seeking to obtain crude 'at cost price' no longer applied.

In order to gain a foothold in the producer countries, the so-called newcomers, that is the independents and the state companies, had to offer terms which had the overall effect of making their oil dearer than the crude 'at cost price' of the already established operators would have been.

That rise in payments to host countries affected the concessions already obtained previously, as well as all the new ones which followed later, so that the importing countries ended up by paying more for their crude than they otherwise would have done.

Naturally, the newcomers tried to gain favour with the producer countries. They were now busy doing what ought to have been done twenty years earlier. Dr Frankel commented that by acting as second-class producers the consumer countries concerned had reduced the power to exert pressure which they would have had if they had become dominant purchasers.[57]

The British expert did not accuse France and Italy (to which Spain may be added) of following an irrational policy with their respective state companies, but he claims that they would have benefited from putting into practice twenty, or even ten, years earlier an attitude which was not advantageous in 1968. One cannot but concur with that view. The problem with the state companies is that they were too late in entering exploration and production abroad.

But one cannot put the clock back. Besides, the case presented by Dr Frankel is in rather broad terms. It is possible to put forward some considerations which show things in a rather better light.

Since 1950–55 profits have never been as high as the level of 70 cents per barrel or more that in some cases was achieved then: they have usually been somewhere in the range of 15 to 30 cents. So there are still profit margins, even if these are smaller. It has usually cost a country less to run a state company that itself performs extraction abroad than to pay foreign companies the prices which include those profit margins.

Even if there is less financial advantage from extracting crude than there was in the past, there are other advantages to consider than profit alone. A high proportion of production costs is paid in the currency of the consumer country, for employees' pay, investments, equipment, building works, transport, etc. It is possible to select the host country according to its present or potential purchases from the consumer country. Economic and political co-operation pacts can be signed when the oil exploration agreements are formalised. The activities of ELF/ERAP/SNPA contributed or saved nearly 200 million dollars in the balance of payments in 1967, which was equivalent to about half the positive balance. France's saving in foreign currency for 1971 as a result of that group was estimated at 350 million dollars. Admittedly there is less advantage when convertible currency is available.

Although the 'newcomers' have caused a rise in the cost of concessions, they have taken a different line in negotiations with producers, and by trying to dispose of surplus production through the open market they have caused reductions in selling prices that have been beneficial to the consumer countries. So long as the big companies dominated the market, the prevailing price usually charged for petroleum products included a sizeable profit. The export policy of the USSR has helped to bring prices down from the levels charged under that rigid tariff but the commercial dynamism of the European independents and state companies has contributed towards minimising prices.

In 1970 there was no longer a sellers' market. The situation had changed since Dr Frankel's discussion of it. European demand was greater than had been originally estimated, shipping costs were much higher until sufficient supertankers became available, marine insurance premiums had been raised and the cuts in the volume of production imposed on large-scale producers in Libyan territory had greatly reduced the tonnages available on the market.

The producer countries found themselves in a position of strength and up to now they have been able to force operators, whether international, private or public enterprises, to give way to pressure.

The state of affairs prevailing in the period 1970–74 cannot last for ever. It is possible to envisage that if the producer countries of North Africa go on making such heavy demands and if they bring the Middle East states in OPEC or OAPEC into line with their attitude, then the companies might react by adopting a concerted strategy of retreat. The companies might band together as purchasers and refuse to act as a buffer between the host countries and the consumer countries. This would amount to the formation of a 'counter-OPEC' by purchasers.

Dr Frankel rightly points out that there was failure to act at the most favourable time in history but he is hard on the consumer countries in that he presents only the negative aspect of their performance in setting up state companies. There has also been a positive result for consumers, from the minor price war that has been waged between the majors and the newcomers. Nowadays the development of a greater degree of consensus than of conflict between the two kinds of company may occur, because they might reach agreement on determined blocking action once they think the producer states are going too far in revenue demands and trying to take over their own natural resources. The companies can move to a natural stronghold already available and fight on behalf of European, Japanese and American consumers.

It is appropriate to consider the advantages and disadvantages for the oil companies of moving towards greater sympathy with consumers. Professor Edith Penrose put the question clearly at a meeting of the Institute of Petroleum in London in spring 1970. She said that the historical motives that had led oil companies to seek their own sources of crude oil were complex. Foreign companies with oil concessions in underdeveloped countries had had to pay a political price for the advantage of controlling their own source of crude oil – that of being stigmatised as exploiters of the host country's natural resources. Why then, asked Dr Penrose, did major oil companies not consider getting out of the production function which had become so much less attractive of late and becoming buyers of crude oil from national companies? The immediate result would be chaos but it was possible to envisage that after a comparatively short transitional period, competition among eager sellers in a situation of over-supply would speedily undermine the edifice of crude oil prices: prices would be forced down much nearer the true cost of production, excluding the government's fixed levy.[58]

According to Dr Penrose's line of argument, an alternative is therefore possible. Yet it is difficult to see how such a change could be arranged. How and by whom would the immense exploitation effort needed to maintain and increase the annual production tonnages be financed? At present the producer countries and their state companies are not in a position to undertake the whole of that task.

The change involved might be painful for the big companies which have been accustomed since the beginning of this century to performing the whole range of oil operations. But it can be more readily envisaged in the case of the state companies of the consumer countries. At any rate, if the producers hold them to ransom too much, the European countries and Japan could change their policy and join together as consumers to strengthen their bargaining power. They could then switch production units that were not sufficiently profitable over to operations rendering support to concerted purchasing power, build up reserve stocks, increase the supertanker fleet, promote production and use of nuclear energy, make all the producer countries compete with each other, and intensify exploration efforts in the North Sea.

The combined effect of all these measures should be to keep price increases in check. An alliance of European purchasers, possessing a central organisation, would certainly be more united than the producers' association, because OPEC has never been able to achieve really consistent co-ordination, given the anarchy in the efforts of the producer states which it represents. Sustained collective pressure should cause producers to exercise discretion, because if the advocates of a hard line failed to compromise in time, then perhaps some less intransigent host countries might separately seek favour with importers, modifying their claims in order to improve relations with customers now in league together in requiring good terms and conditions of sale.

Here again the pattern of relationships between the players in the game comes into operation, so that none of the contenders can achieve complete and lasting victory over another. It is difficult to see the future in terms of a decisive rift or a sudden violent change. The American philosopher C. Lindblom has described as 'incrementalism' the modern process of making changes by a sequence of small stages. He has pointed out that, 'Usually – though not always – what is feasible politically is policy only incrementally, or marginally, different from existing policies. Drastically different policies are beyond the pale.'[59]

Taking into consideration the circumstances of the state companies in the countries with Mediterranean coastlines, the features favouring them and certain weaknesses characteristic of them, one can attempt to suggest future developments. In the competition bringing them into simultaneous alliance and opposition with states and with large companies, the state companies can survive if their functioning is not inflexible but adjusts to the conditions of the moment and if they reach understanding with each other on the formation of a clear link between consumers and producers.

If a state company seeks to play an effective part in competitive action it must become flexible and get rid of much of the rigidity traditionally present in a public enterprise. As J.E. Hartshorn, Industrial Editor of *The Economist,* said at a seminar in Riyadh in April 1967, national oil companies are companies acting as government agents, whether or not they are state-owned.[60] In other words, just as in a capitalist enterprise, where there is nowadays increasing separation between ownership and management, it is not of fundamental importance who owns the company. What matters is the strategy employed by an enterprise. It is possible for the operations of a public enterprise to be fairly independent of the State to which it belongs. On the other hand, a private enterprise can run its affairs on the lines desired by government strategy.

The future of state companies depends on the answer to three basic questions. Will they be able to keep to a difficult path which means avoiding both 'economism' (the testing of theories of profitability) and giving political considerations priority, irrespective of the demands of economic realities? Will it be possible for the state companies of the producer countries and those of the consumer countries to co-operate in such a way that the latter companies bring the former companies into production, refining, transport and distribution without causing too great a rise in prices, and will the two sides be able to agree on sharing the results of co-operation equitably? Finally, in the industrialisation which is bound to occur in the countries of the third world during the present decade, will the state companies show enough dynamism and flexibility to be able to undertake new forms of collaboration without ruin to themselves and to manage to give substantial help to the country concerned, as ENI did during the decolonialisation period?

The survival of the state companies depends on how they cope with the matters involved in those three questions. If they do not succeed in giving themselves a dynamic image that is sufficiently distinct from that of the majors, then they are doomed to becoming pale reflections of the big companies, doing the same thing as them but with narrower scope and less efficiency.

Notes

[1] According to *Algérie-Actualité,* 4–10 January 1970, p. 4, the turnover was 377 million dinars for 1966 and 1,468 million dinars for 1968. The profit for 1968 was given as 292 million dinars.

[2] Statement by the Chairman of Sonatrach, Mr Ghozali, reported in *BIP,* 1677, September 1970, p. 3.

[3] 'The 200 largest industrials outside the US', *Fortune,* August 1970.

[4] Royal Dutch Shell in 1st place, 9,738,410 million dollars turnover, 173,000 employees. BP in 5th place, 3,424,080 million dollars turnover, 68,000 employees. Petrofina in 57th place, 1,053,640 million dollars turnover, 19,900 employees.

[5] ELF, *Report of the Management, 1969,* p. 12. It is important not to confuse Entreprise de Recherches et d'Activités Pétrolières (ELF/ERAP) and the ELF group, comprising ELF/ERAP plus all the affiliated companies.

[6] *Entreprise,* 791, November 1970, p. 109. ELF comes in third place

in size of profit made by French companies in 1969. CFP heads this list, with 718,546,000 francs.

[7] ENI, *Financial Year 1968,* summary of the Report of the Board of Directors, p. 33.

[8] Consolidated balance sheet for the group as at 31 December 1968 and 31 December 1969.

[9] ENI, *Financial Year 1968,* p. 10. SNPA produced 6·3 thousand million cubic metres in 1968, 6·5 thousand million in 1969 and about 7 thousand million in 1971.

[10] P.H. Frankel, *Mattei: Oil and Power Politics,* Faber and Faber, London 1966, p. 137.

[11] *ENI Informazioni,* 14, 5 August 1969, p. 1. See also *Petroleum Intelligence Weekly,* 14 April 1969, p. 4, reporting that the group would probably achieve self-sufficiency in crude in 1970. In fact, the 1972 production was probably some 15,000,000 metric tons. The agreements made with Libya in 1972 were expected to produce rapid improvement from that level.

[12] *Petroleum Press Service,* September 1970, p. 349.

[13] ELF, *Report of the Management, 1969,* p. 20. Without SNPA, production was 22 million metric tons. In 1970, without SNPA, the total increased to 23·4 million metric tons.

[14] ELF, *Monthly News Bulletin,* February 1972, p. 9.

[15] ELF, *Report of the Management, 1969,* p. 63.

[16] ENI, *Financial Year 1968,* summary of the Report of the Board of Directors, p. 13.

[17] Ibid., p. 9.

[18] *Petroleum Press Service,* May 1970, p. 167. CFP had a 15 per cent holding, which has since become 13·34 per cent. ELF, which had a 2·8 per cent holding, managed to obtain an increase to 10 per cent, in line with the amount drawn off for its refineries. Investment estimates for trebling the capacity of the South European Pipeline, PLSE, were 900 million francs. See also ELF, *Monthly News Bulletin,* 25 September 1970, p. 7.

[19] Schedules XI and XIbis to the Accords of July 1965.

[20] D. Murat, *L'intervention de l'Etat dans le secteur pétrolier en France,* op.cit., pp. 33 and 42.

[21] ENI, *Financial Year 1968,* p. 13.

[22] Ibid., p. 17.

[23] On 31 December 1969 the ELF group had a fleet of 15 vessels, with a total tonnage of 935,000 tons, 12 of them on time charter. This chartering is mainly from two companies, SFTP, in which the State is a shareholder, and CNN, in which ELF has a 35 per cent holding. The

position of CFP is different: through an affiliate in which it has a 99 per cent holding, Compagnie Navale des Pétroles, which has shipping capacity of 1,300,000 tons and 16 vessels, CFP has the largest shipping capacity of any French enterprise.

[24] ELF, *Report of the Management, 1969,* p. 34.

[25] Statement by J. Chenevier, Chairman of BP in France, at the launching of the *Blois,* a 240,000-ton oil tanker, at the end of 1970.

[26] A. López Muñoz and J.L. García Delgado, *Crecimiento y crisis del capitalismo español,* Ed. Cuadernos para el diálogo, Madrid 1968, p. 163.

[27] This is the conclusion reached after lengthy consideration by R. Tamames, *Estructura económica de España,* Ed. Zyx, Madrid 1964, pp. 283 et seq.

[28] Order 69,107, *Journal officiel de la République Algérienne* (Algerian official gazette), 31 December 1969, pp. 1262 et seq.

[29] Section 29 of the Order.

[30] Section 30.

[31] Section 35.

[32] Section 39.

[33] Section 37.

[34] Quoted in J. Baumier, *Les maîtres du pétrole,* op.cit., pp. 139–40.

[35] *Stampa e oro nero,* 35 volumes, Gatto Selvatico, Rome 1963. The film *The Mattei Affair,* first shown in 1972, succeeded in emphasising this psychological battle and its strongly nationalistic nature.

[36] *Corriere della Sera,* 13–17 July 1962.

[37] *Corriere della Sera,* 27 July 1962.

[38] Ibid., 27 July 1962.

[39] Ibid., 27 July 1962.

[40] Ibid., 30 August 1962.

[41] J.K. Galbraith, *The New Industrial State,* Hamish Hamilton, London 1967, p. 101. For example, the State has a holding of approximately 50 per cent in BP but BP operates in a way which differs very little from the way in which a wholly private enterprise functions. In May 1965, Harold Wilson, as Prime Minister, told the House of Commons that, 'Although reports (from the Government appointed directors on the board of BP) are quite frequent, they are mainly of an oral nature . . . and mainly informal'. (*Hansard,* vol. 710, cols. 1111–13). See also P.R. Odell, *Oil: the new commanding height,* Fabian Society, Fabian Research Series no. 251, London 1965, p. 5.

[42] *Rapport sur les entreprises publiques* (known as Nora report), Documentation française, Paris 1967, pp. 17–18.

[43] Ibid., p. 18.

[44] Ibid., p. 25.
[45] ELF, *Report of the Management,* 1969, pp. 58–9.
[46] M.V. Posner and J.J. Woolf, *Italian public enterprise,* Gerald Duckworth and Co., London 1967, p. 107.
[47] *Petroleum Press Service,* August 1968, pp. 295, 300 and 301. The US group under reference in this study by the Energy Division of the Chase Manhattan Bank was the five American majors and the leading independents, i.e. 28 companies which in 1967 together accounted for more than half the company investment in the world, excluding the Communist bloc. The financial structure of these 28 companies is analysed each year in a report. See Chase Manhattan Bank, *Annual Financial Analysis of a Group of Petroleum Companies.*
[48] *Petroleum Press Service,* April 1970, p. 131.
[49] J. Huré, *Quelques élèments de base en vue d'une réflexion sur la politique française du pétrole,* paper given on 24 January 1966 at Institut des hautes études de la défense nationale.
[50] P.H. Frankel, *L'économie pétrolière. Structure d'une industrie,* Librairie de Médicis, Paris 1948, p. 199.
[51] The French Government preferred to negotiate alone with the Algerians and expected the companies then to find ways of preserving at least a minimum of co-operation in oil operations in the Sahara.
[52] P.H. Frankel, *Mattei: Oil and Power Politics,* op.cit., pp. 149 et seq.
[53] P.H. Frankel, op.cit., p. 159.
[54] G. Petrilli, 'L'éthique de l'entreprise publique', *Synopsis,* November-December 1969, p. 37.
[55] P.H. Frankel, 'The current state of world oil', *Middle East Economic Survey,* XI (45), 6 September 1968.
[56] BIP, 1171, Documents, p. 13.
[57] Ibid.
[58] E.T. Penrose, 'Is Integration Worth While?', *Petroleum Press Service,* June 1970, pp. 203–4.
[59] C.E. Lindblom, *The policy-making process,* Foundations of Modern Political Science series, Robert A. Dahl (ed.), Prentice-Hall, Inc., Englewood Cliffs, New Jersey 1968, p. 26.
[60] Quoted in E.T. Penrose, *The Large International Firm in Developing Countries: The International Petroleum Industry,* George Allen and Unwin Ltd., London 1968, p. 219.

5 The Death of Liberalism in Oil Affairs

The setting up of state companies in the countries on either side of the Western Mediterranean is only a particular instance of a more general phenomenon. The entry of the State into economic competition as an entrepreneur in the oil business is a recent manifestation in one sector of the economy of a wider, more general and much older policy. Joining in the mêlée may be only a small measure compared with the long and difficult task of general regulation of affairs. The direct intervention of governments in oil operations should not be allowed to obscure the much more important, although less conspicuous, work performed indirectly through special legislation, detailed regulations, decrees, orders and instructions. The whole of a line of activity, whether in the hands of private or public, national or foreign interests is affected by indirect action. In that context the State is not in the position of a participant in the oil business and as such theoretically subject to the same rules of competition as the other operators. It places itself slightly apart from the economic game which it is witnessing, assuming the function of setting the rules and watching to make suare that the various players obey those rules. It endeavours to so arrange matters that those engaged in the occupation accept (positively or negatively, as the case may be) the requirements which have been imposed by means of the exercise of its authority.

A fairly convincing case has been made out to show that the entry of the State into the economic arena means that it did not manage soon enough and consistently enough to devise an appropriate oil policy and obtain compliance with this. It is, after all, one of the most important duties of the administrative authorities of a country to make known a set of standards to which firms must conform if they wish to operate in the national territory. In practice this requires that the relevant political organs determine, arrange and impose a system within which economic activities may be conducted.

This conception of the matter is only the expression in simple terms of the principle that the essential aims of economics should be governed by politics. The reverse is not true. Accordingly, a country's leaders must possess the political aptitude that is needed in order to undertake the combination of planning, ideas and decisions from which a policy is formed.

The countries with Western Mediterranean coastlines provide an especially interesting illustration of state interventionism in oil affairs. Starting from the sources of production in their control the producer countries of North Africa wish to extend their rule to the various stages of processing and handling of petroleum products, by means of a combination of a certain amount of voluntarism and a large measure of constraint. The consumer countries on the other side of the sea are dealing with the sequence of stages for the products in the opposite order: since final use of the products takes place in their territories, they are endeavouring to exercise strict control over the range of preceding operations by applying powers of authority whilst respecting the initiative of individuals and corporate undertakings. It takes time to introduce measures dealing with all the stages from the well to the pump or from the pump to the well. What the European countries accomplished earlier in regulating the activity of the large international companies the countries of North Africa are now trying to achieve in their own way for themselves in connection with the foreign companies, some of which are of European origin.

One country on either side, Algeria and Spain, may be taken as illustrative of the course of developments.

Under the agreements made at Evian, Algeria had to be content with taking a share in exploitation of the riches of the Sahara and respecting the rights that had previously been acquired by operators. But in the course of time that country's ambition of taking control of the national oil resources was displayed more and more openly. Radical measures affecting the non-French companies were decided on. Government regulations imposed more severe constraints on the French companies. From 1969 onwards it was clear that Algeria was seeking complete 'recovery' of its sources of production, in order to obtain full possession of the proceeds from them, for the purpose of financing very large investments under the first Four Year Plan (1970–73).

Algeria undoubtedly provided a pointer for the group of producer countries, most of which were very late in becoming aware of their bargaining power in dealing with the oil operators. But it is symptomatic that even in accelerating matters as much as possible Algeria has to proceed by stages and cannot afford to go too quickly. Even after the nationalisations of 1971 Algeria still needs to keep the French companies as minority partners.

A certain likeness may be seen in the development of political awareness in a consumer country with little domestic production, Spain, which has now entered international competition. Initially Spanish ventures in oil had neither the financial nor the international scope to permit political intervention at all stages of oil operations. The international structure of oil affairs imposes constraints on any attempts at comprehensive action, that is, measures dealing with all stages of the product until it reaches the final consumer. Because of the impossibility of doing everything at the same time, it was decided to start from the point where endeavours could be most remunerative – the final stage. Accordingly, the first objective was to act at consumer level by removing supply to the user from the orbit of purely private interests. For that purpose an import monopoly was created in 1927. Next, moving on to the next stage upstream, the plan involved introducing refining facilities, with the long-term aspiration of importing crude only, not products already finished by other processers. Again with a view to escaping high costs and reducing the outflow of currency, the next project was to obtain so far as possible the shipment of crude from the country of export by a fleet under the national flag and not by foreign shippers. Because periods of shortage had been experienced, a project further upstream than the previous endeavour was taken up, namely that of systematically prospecting the national subsoil in an attempt to discover worthwhile deposits that would reduce dependence on foreigners for the means of creating energy. But since success in that procedure seemed to be distant in time, costly and chancy, it was considered necessary to engage in efforts abroad simultaneously, by taking exploration rights and exploitation concessions. The purpose in doing so was to hold command of some wells abroad in order to be able to exercise greater control over the product which arrived at pumps in Spain.

Thus, over a period of forty years Spain gradually built up the system occurring in any interventionist policy of a country lacking adequate oil resources of its own.

This chapter is concerned with the fact that the Western Mediterranean countries are basically interventionist in oil affairs, whether they are producers or consumers. Here again it is proposed to continue to apply a critical approach and to attempt to assess the merits of such a policy.

The United States is often invoked as an example to show the value of free competition in oil and the benefits to a national economy of a policy of non-intervention by the public authorities. That argument overlooks the fact that US liberalism in oil affairs is strictly for export and the authorities take care not to apply it in connection with the domestic market. The United States is the only country other than the USSR to be both a large producer and a large consumer.

During the 1960s the US Federal Government strove to isolate the domestic market from the rest of the world. The motives for this overall protectionism were partly strategic, as the reserves were equivalent to ten years' production, or a few years longer when Alaskan oil was included. There were also technical reasons, in that the productivity of the wells was low and there was a quota system for production. Finally, there were economic considerations, the operating costs for most of the wells being very high. In view of the absence of real competition from abroad, the USA has been described as a paradise for small and medium-sized oil producers. In order to penetrate beyond this isolationist wall protecting the various interests importers had to pay a high entrance fee to the administration.

A very high price has been paid for this security. According to various estimates, in 1968 this policy cost American consumers between 10 and 30 thousand million francs. Enriched by the large profit margins obtainable from concentration of refining and from distribution in such a vast market and assisted by the 'depletion allowance' whereby a proportion of receipts may be deducted from taxable income, some twenty companies, five of which are among the majors, were easily able to finance from their income from the American market their own expansion throughout the world, and especially in places where profits were still large (Venezuela, the Middle East, Libya). In 1968 the American majors accounted for 27·5 per cent of world sales of petroleum products and 40 per cent of sales outside the United States, excluding those in the Communist bloc. The aggregate for the American oil companies represented between 50 and 60 per cent of world oil production in that year, including the Communist bloc.[1]

This preponderance of the American companies at world level derived from their predominance in holdings in sources of production, because now, as at the beginning of the history of oil, the operators holding the wells collect the profits and make a great deal more money than less fortunate oil enterprises. The net annual profit from American oil ventures abroad for the years before 1969 was some 1·5 thousand million dollars.

American liberalism in oil affairs may, therefore, be much vaunted in official statements and in ideological discussion but a closer examination of the facts shows that free competition is strongly supported abroad and little practised at home.

In fact, no country which supplies or purchases a raw material so vital to economic development can be a passive spectator of oil extraction and processing. That principle is especially evident in connection with the geographical area under special consideration in this book. In that region state involvement may be said in many respects to be greater than elsewhere, although less thoroughgoing than in the Communist countries where it is an article of dogma. But is this the best policy? Would it not be possible to trust the multinational enterprises instead of constantly imposing restrictions on them or promoting in order to compete with them public enterprises somewhat resembling Trojan horses introduced into the business by governments? It is therefore appropriate to weigh carefully the advantages and also the overall cost of this interventionism practised by all the producer countries and consumer countries of the Western Mediterranean.

The producer countries of North Africa

The producer countries possess the raw material. It is natural that in awareness of the potential wealth involved these states should be quite strongly interventionist. Algeria and Libya, in differing ways, are implementing one of their principal objectives, namely that of increasing the taxation imposed on the operating companies. Algeria has taken a stronger hold over oil produced in the national territory than Libya has, doing so in order to make it assist the nation's economic development. In the conditions of a sellers' market, the sellers find that an interventionist policy has certain advantages, although there are some disadvantages.

The upheaval in 1970, producing a very great increase in demand, reduced flexibility of supply and aggravation of the shortage of shipping, brought a change in circumstances in favour of producer countries. They were quick to take up the opportunity, by asserting new claims as early as December 1970, at the OPEC Conference in Caracas.[2]

During 1970 Libya managed to obtain a sizeable rise in the posted prices and in the rate of tax applicable to them (increase of 30 cents for light crudes and still more for the heavier oils, together with a rise of 2 cents a year for five years, also a gradual increase in the profits tax, raising it to 54–5 per cent by 1971). But this was only a prelude to the worldwide wave of bargaining which started when the Persian Gulf states joined Iran in following Libya's lead.

New moves at that time took place in three stages, at Teheran, Tripoli and Algiers in the first few months of 1971. These conferences yielded changes in prices and taxation and some alterations in conditions for oil operations which will be mentioned later.

The financial terms of the Teheran agreement[3] brought the total revenue per barrel to 1·30 dollars in 1971 and provided for subsequent rises to lift the level to 1·50 dollars per barrel in 1975. Supplementary payments to the governments of the region were to total 1·2 thousand million dollars in 1971, rising by stages to 3 thousand million dollars in 1975. However, the countries concerned undertook to refrain from new demands for a period of five years.

The increases were spread over a period of time and involved the following:

1 A uniform rise of 35 cents per barrel in the posted prices of crude at the Persian Gulf terminals. Of this amount 2 cents were to cover differences in transport costs.
2 A special increase of 5 cents per barrel in the posted prices 'to reflect the growth in demand for crude during the term of the agreements'.
3 An annual increase of 2·5 per cent in the posted prices, to cover the rate of inflation for goods imported by the Persian Gulf states.
4 An increase of a few cents per barrel under a new system for dealing with 'differences in gravity'.
5 Taxation of companies' profits was to be stabilised at the rate of 55 per cent.

Accordingly, Resolution 120, adopted by the members of OPEC at the Caracas Conference at the end of 1970, was put into effect almost unmodified for the countries exporting their oil via the Persian Gulf.

The companies had hoped that at Teheran they would be able to make an overall agreement covering both the Persian Gulf and the countries of North Africa. But because of the demands of the Libyan and Algerian Governments, which sought to obtain special terms for the countries close to industrial Europe, a fresh set of negotiations in Tripoli followed those in Teheran. At those negotiations Libya was the chief spokesman for production west of Suez (Algeria and Libya) and for that part of the Eastern oil delivered to the ports of the Eastern Mediterranean (80 million metric tons, including 25 from Saudi Arabia and 55 from Iraq).

After several breakdowns and after threats from each side, an agreement was finally signed on 4 April 1971. This agreement was valid for five years but commentators pointed out that in the first round, in September 1970, Libya had already undertaken to refrain for several years from seeking to change the terms of the bargain. By the Tripoli agreement the posted price was raised from 2·55 dollars per barrel to 3·45 dollars and there were to be annual increases.[4] But the companies obtained the condition that this price included a rising geographical allowance starting at 0·25 dollars. This allowance was to be reduced, or perhaps even removed altogether, in the event of re-opening of the Suez Canal or of large decreases from the artificially high freight rates prevailing at the beginning of 1971. The making of such a condition created a slight deviation from uniformity between the two main oil producing regions, enabling greater competition to be introduced between those suppliers of Europe west of Suez and those east of Suez.

The claims in connection with conditions for oil operations (obligatory reinvestment in the host country, assumption of partial control) which Libya put forward during the negotiations had completely disappeared in the final agreement, leaving only a vague reference to the companies' obligation to set aside an adequate quota of their profits for exploration for new deposits in the host country. This provision could be interpreted fairly subjectively by one or other of the sides and gave Libya an opening for making further claims before the end of the five-year period, as the situation has remained favourable to producers.

Algeria was able to gain in two ways, namely in revenue and in conditions. It obtained alignment of its posted prices with those negotiated in Tripoli. It can claim greater proximity to industrial Europe but its oilfields are much further inland than those of Libya.

In the event, the special feature in Algeria was a more spectacular move which destroyed at a blow the judiciously designed arrangements made under the 1965 treaty.

It has already been mentioned that one of the primary aims of Algerian oil policy was to 'recover' basic control of its oil resources. Beginning from the stakes inherited from the time of French administration of the country, Algeria rapidly enlarged its sphere of control. The first act was the complete takeover of non-French interests. This had been followed by an attack on the French interests which had earlier represented two-thirds of production. Between the Teheran Conference and the Tripoli Conference matters had reached the point of breakdown of the interminable Franco-Algerian negotiations and the outcome of the Teheran Conference was awaited. On 24 February 1971 President Boumedienne announced 'the entry of the Revolution into the oil sector' by means of a set of spectacular measures – taking a majority holding (51 per cent) in the French companies operating in Algeria along the lines of the Getty-Sonatrach agreement, appropriation of the whole of the pipeline system, complete nationalisation of all natural gas resources. The Algerian Head of State said on the same occasion that his country would continue to supply the French market and he undertook to provide compensation to the French companies 'on the basis of that paid to the international companies earlier'.

Algeria had been expected to one day take over part of the oil resources in French hands but there was uncertainty about the timing of the move. In the event, Algeria probably preferred to strike quickly with a heavy blow, taking advantage of a favourable world situation and knowing that the French had less scope for reprisal than the British and American groups. The violence of the action might also have been due to fear of finding that Libya had joined the moderate camp, thus putting Algeria at variance with its earlier pronouncements.

However, all the plans made in 1971 under the Teheran, Tripoli and Algiers agreements were overturned by the October Revolution of 1973, during the third Arab-Israeli war. This important development will be studied separately in Chapter 7. For the moment it seems relevant to mention that it followed a process of world inflation which no industrialised nation managed to control. After the financial measures taken by President Nixon in 1971 and 1973 the dollar fell sharply in relation to other strong currencies. As oil contracts usually quote terms in dollars, the depreciation of US currency caused risk of serious effects on supplier countries. By maintaining solidarity of action among themselves, those countries were able until June 1973 to partially offset the erosion caused by crude prices when they were already suffering from the inflation prevailing throughout the world. They also obtained adjustment and guarantees in connection with the declining value of the dollar, achieving better results by applying an index based on a mean overall rate for the richest countries of the world.

But the crisis of autumn 1973 produced an opportunity for complete revision of world oil arrangements. There was a sharp change in ownership proportions, with the signature in 1974 by the Persian Gulf states and Libya of agreements giving them a 60 per cent holding. A much more gradual policy had been provided for earlier, setting the stake at 25 per cent in 1973, rising by instalments to reach 51 per cent in 1982. Great differences from the past are to be expected if the producer states find themselves in control of larger quantities of crude than those obtained by the companies under the concessions which they held prior to the change.

In posted prices and revenue there has been a huge jump, making the levels in January 1974 four or five times those of January 1972, as table 4 shows.

Interventionism is justified if it causes an increase in revenue and facilitates a policy of expansion.

Before the increases of 1970–71 Libya was already obtaining higher

Table 4

Posted or tax reference prices (dollars per barrel)

	20 Jan. 1972	1 Jan. 1973	16 Oct. 1973	1 Jan. 1974
Persian Gulf				
Arabian Light (34°)	2·479	2·591	5·119	11·651
Abu Dhabi Murban (39°)	2·540	2·654	6·045	12·630
Mediterranean and Africa				
Arabian Light (34°)	3·370	3·451	7·149	13·647
Libyan (40°)	3·673	3·777	8·925[a]	15·768
Nigerian (34°)	3·446[b]	3·561	8·310	14·691
Venezuela				
Oficina (35°)	3·261	3·477	7·802[c]	14·247[d]

[a] 19 October
[b] 15 January
[c] 1 November
[d] Excluding sulphur premium

Source: *The Petroleum Economist,* February 1974, p. 42

revenue per barrel than any of the Middle East states. The only country in the world with a higher rate than Libya was Venezuela. For 1969 Libya's total oil revenue was 1,132 million dollars. Algeria was far below Libya in this respect, obtaining in that year 255 million dollars. These large receipts were due to clever bargaining by the negotiators of the two countries but also resulted from proximity to the European markets at a time of shortage of shipping.

Since 1971, and even more so since 1973, there has been a remarkable rise in the receipts of the North African countries. For 1974 the World Bank estimated Algerian receipts from hydrocarbons as 3,700 million dollars, compared with 1,000 for 1973. Consequently the 1974 budget was 33·4 per cent higher than the 1973 one, at 14,173 million dinars. According to the Minister of Finance, Mr. Smail Mabroug, the capital budget had risen consistently from 1971 to 1974, the proportions being 23, 33, 44 and 55 per cent respectively.

According to the World Bank, Libya's oil revenue was expected to rise from 2,200 million dollars for 1973 to 8,000 million dollars for 1974, having been 1,295 million dollars in 1971.

A further advantage of interventionism is that it enables hydrocarbons to be used as a lever to speed up development of the national economy. According to the Libyan Minister for the National Plan, in 1967 oil already represented 60·1 per cent of the Gross National Product. In 1970 some 99 per cent of government revenue came from oil. For Algeria, which is a moderate oil producer and a country where modernisation is further advanced than in Libya, the proportion provided by oil was smaller, accounting in 1970 for only between 16 per cent and 16·5 per cent of the GNP and only 23 per cent of government revenue.[5]

Yet Algeria has shown itself to be the more anxious of the two countries to make oil serve national development. It might be argued that the prolonged crisis in Franco-Algerian relations which arose in 1970 was a way of leading up to drastic action in order to obtain sources of funds for carrying out the ambitious Four Year Plan.

The first Four Year Plan provided for investments of more than 28 thousand million dinars, which meant a rate of 6·5 thousand million dinars a year. Of that total amount industry received 12,400 million, i.e. 45 per cent, and agriculture 4,140 million, i.e. 15 per cent.[6]

Hydrocarbons play an ambiguous part in a national economy. They are the means of getting the development of the country moving and at the same time the part of the economy requiring massive amounts of capital for its own development. Thus, they are a means of giving impetus to the economy, yet it is hoped that they themselves will be made profitable by industrial development.

Under the second Four Year Plan (1974–77) investments were to be not less than 70 to 80 thousand million dinars. In a speech made in Algiers, President Boumedienne even suggested a figure of 100 thousand million dinars.

From the preparations for that Plan it was possible only to discover the guiding principles and not to learn by what means they were to be implemented.

The directive finally adopted on 14 May for the guidance of all those responsible for implementation of the Plan stated that 'the second Algerian Four Year Plan (1974–77), similarly to the first Plan (1970–73) is inspired by the fundamental aims of the country, seeking to build a socialist society and bring all citizens the benefit of the fruits of economic development and social progress'.

The general objectives of the Algerian plan were 'consolidation and enlargement of the bases of economic, social and cultural development laid by the achievements of the Algerian Plans, namely the Three Year

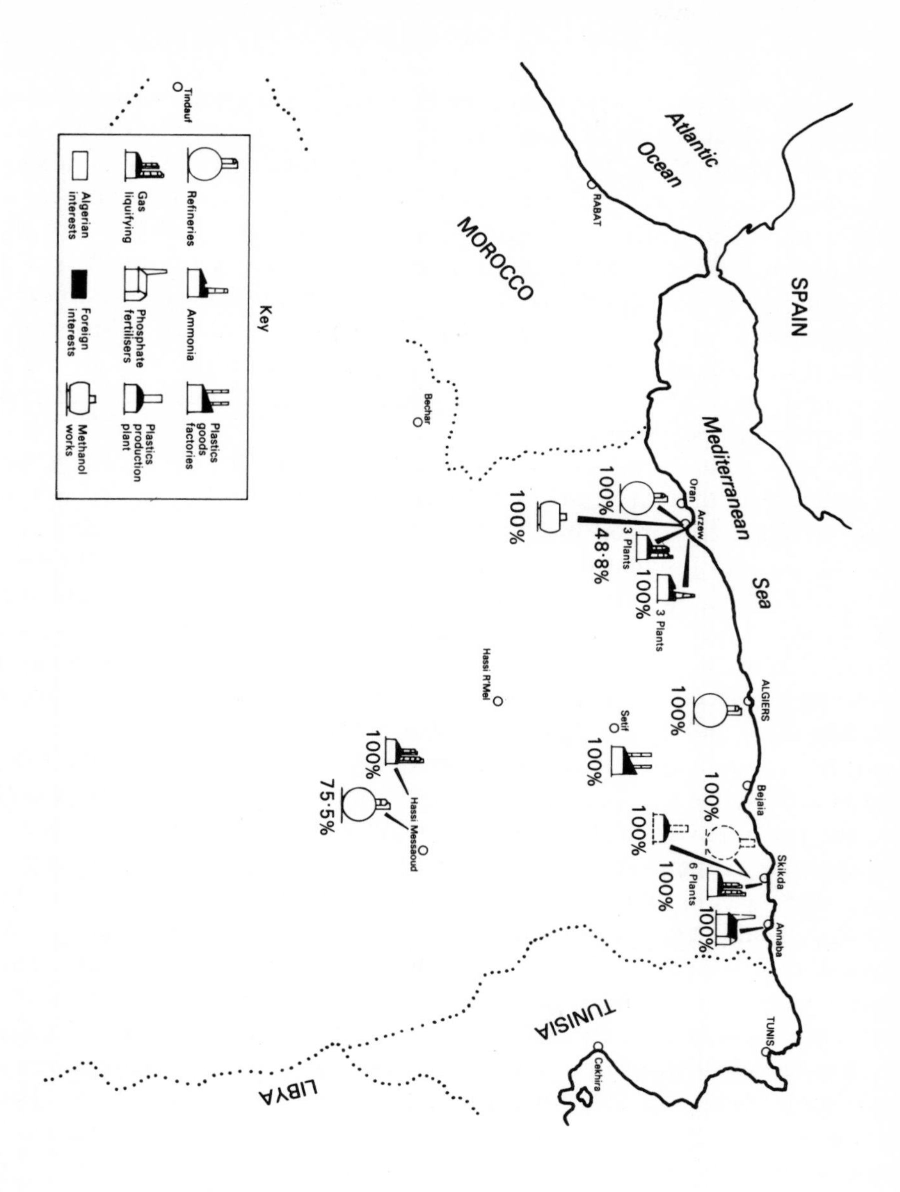

Fig. 19 Refining and petrochemicals production in Algeria, 1972
Source: *Revue Française de l'Energie,* 255, July–August 1973, p. 410

Plan (1967–69) and the Four Year Plan (1970–73), those bases being indispensable to the correct fulfilment of the needs of the people, to causing the full potential of Algerian society to flourish and to the promotion of Man'.

In the second Four Year Plan greater priority was given to 'agricultural development secured by putting into effect the agrarian revolution'.

Despite the special attention given to agriculture there are grounds for wondering whether it will not continue to be treated as a poor relation, as in the preceding plan, although the overwhelming majority of the population is rural. If the oil sector is given 'the priority of priorities' will it become a sector apart, submerged by a traditional socio-economic agglomeration, or will it become a pilot sector capable to infusing the whole of the social apparatus with a spirit of modernity? The answer is not clear if one bears in mind the fairly serious risks attendant on pursuing interventionism on a large scale.

Any state voluntarism is accompanied by risks. The adoption of any policy suggests that there are definite alternatives. Clearly, the decision to place the emphasis on oil means that other efforts are to be subordinate. But the present fascination with that raw material will not last for ever. What will be the position in twenty or thirty years' time of countries which have staked everything on oil? It is probably because of recognition of that future risk that the Algerians and the Libyans have shown concern for being involved not only in exploiting resources but also in transporting and refining it and in investment in petrochemicals.

In the remainder of the era of the predominance of oil production there is another risk for the countries of North Africa. That risk is linked with the one just mentioned. Oil companies are not in the habit of behaving as philanthropic organisations: their investments are planned according to their own overall benefit and the scope is not confined to any one country. Their strategy is worldwide. Their possibilities of action extend beyond national governments which imagine themselves able to obtain passive obedience. Already in Libya it has been found that companies have been ceasing to carry out exploration of sites which were promising but too far away from the present locations of operations. Spokesmen of the régime have regularly denounced this selective 'pillage'. Even more serious is the fact that since the negotiations of 1970–71 a net decline in exploration has been observed; the companies have preferred to concentrate on accelerated exploitation of deposits already discovered, without concerning themselves overmuch with investigating new reserves.

Finally, the oil business is not the miraculous solution to all problems which the leaders of the most imaginative producer countries have tended to expect it to be. It may instead be a powerful cause of economic and social decline, because oil makes the development of a country come from the top of the economy and not from the foundations. This is particular true of Libya. In a carefully documented article, an Oxford economist, Robert Mabro, has described Libya as showing, because of oil, the characteristics of a state receiving income without providing labour input.[7] He observed shrewdly that:

> The oil states cannot be classed with either the industrialised nations or the underprivileged countries of the Third World. Unlike an industrialised country, which itself achieves the resources required for consumption, investment and trade, a state receiving income without providing labour input obtains its funds without supplying labour in order to obtain that income. Unlike the poor countries, a state receiving income without providing labour input enjoys certain fruits of development – constant raising of the standard of living and availability of abundant capital for investment in production and in improving social conditions. It is solely in order to be able to gather those same fruits that the countries of the Third World make efforts, plan, call for more aid and strive, often in vain, to make greater savings. Development is both a system and a result. Libya needs to discover the system but is already installed at one end of the system and some observers consider, perhaps rightly, that she has arrived there too quickly.[8]

Despite the prodigious growth of the oil industry in Libya, Mabro points out two phenomena which are necessarily disquieting. Firstly there is the contrast between the shortage of funds in the non-oil part of the economy and the abundant receipts obtained from the export of a raw material. There are large quantities of capital available in an economy where many sectors are far short of modernity. Secondly the distribution of money in the population is not governed by productive labour. Funds are obtained without working to achieve them. In a normal economy income is linked with employment in a productive process but in a state where labour input is not contributed money is received by the Government and distributed to households, administrative bodies and enterprises in the form of disguised gifts, not being supplied in return for a positive contribution of labour. (This arrangement is preferable to the enrichment of certain favoured individuals and undertakings only, as has occurred in many Persian Gulf states). It is difficult to get the indigenous population to perform productive work when there is a constant flow of money into the national Treasury despite little active participation by people living under this system which supplies ample wealth.

The consequences for the structure of the Libyan economy are fairly noticeable. Agriculture is not advancing: worse still, it does not even supply enough to meet the country's needs. In 1952 it supplied sufficient food for the subsistence of 1,200,000 inhabitants with one of the lowest levels of consumption in the world. In 1967 Libya imported more food than it produced (25 million Libyan pounds spent on imported food; national food production worth 20 million Libyan pounds). There is a growing new proletariat in the urban areas but oil cannot employ more than 4 to 5 per cent of the labour force and a shortage of skilled labour is all too evident. Industry is scarcely developing. However, the third sector (building, transport and services) is growing out of all proportion to the other sectors or national needs. Ownership of real estate which yields profits is more advantageous than driving a heavy vehicle or being employed on an industrial assembly line. In Libya there is now one vehicle per twenty inhabitants, which puts that country at the head of the list for ownership of vehicles in Africa. In sum, oil is not producing the traditional sequence of development from agriculture to industry and then services. It is reversing the order and promoting above all the growth of the third sector.

Because of these basic phenomena the weaknesses of Libyan development are manifest. 'The paradox of development is that it presupposes starting from some degree of initial development', Mabro continues. Wealth is not always a substitute. Oil is an extreme example of that paradox. Because of its almost limitless industrial potential it might be thought to be a good starting point. In fact, it is the potential which makes it so valuable. And that is what makes it difficult to deal with; the technical skill, the knowledge and the organisational methods required to manage this very versatile substance profitably are too complicated for a backward economy. A state benefiting from this commodity but not contributing labour to industrial production is not able to put its endeavours straight away into oil, this miraculous substance from which its income is obtained. It begins by putting expansion efforts into the wrong end of the economy, the third sector. It is to be hoped that the use of services and the investment in improving social conditions may in time lead to enhancement of the value of human resources, so that the conditions for development that were lacking at the outset may be attained. Only then will a state unaccustomed to labour-intensive industry be able to make the oil industry and trade an integrated part of its economy. The judicious solution of the problem consists of first using oil income for training personnel and then treating it as a result of a production process.[9]

Fairly thorough analysis of this state of underdevelopment in a country which is both poor and at the same time well-favoured by fortune shows that an unsatisfactory situation may persist even after revenue per barrel has increased. The Algerians have deduced that change from possessing oil income to making oil play a part in industrial production is achieved by state takeover of national oil resources and operations. But that method of procedure is likely to provoke the anger of the consumer countries, which have for a long time been anxious not to remain passive in the matter of a commodity so vital to each of them.

Interventionism of the consumer countries

After the Second World War certain problems confronted each of the countries of Western Europe. They all had to rehabilitate a national economy ruined by the war, find reliable and reasonably cheap sources of energy and acquire some independence of the country which had emerged as the strongest after the conflict, the United States. For several years it was possible to believe that the restoration of full production in coalmines and modernisation there could serve as the basis for the assured energy supply at acceptable cost which was required as the key to industrial recovery. But European coal soon proved to be incapable of winning the contest with its rival, oil. Europe had only marginal production of the latter commodity and surely it was risking relying heavily on foreign suppliers who would be quick to take advantage of weaknesses on the other side? The State was invested with the leadership in ensuring the success of industrial development and could not postpone action in this vital matter. Italy, Spain and France did not act differently in this connection from other European countries. The only difference was that state interventionism was more vigorous and more noticeable there than elsewhere. The control system that had been designed by legislation between the world wars was revitalised and greater pressure was put on the big companies to comply with it.

It seems to me that Southern Europe is a consumer zone unlike others in that a special degree of *dirigisme* in oil affairs occurs in the endeavour to prevent the status of consumer declining into the status of mere purchaser. It is true that imports have risen enormously but refining capacity has kept pace with them. The authorities have made room for public oil enterprises. The point requiring investigation is the overall cost of this policy.

From figure 20 it may be seen that oil imports by the six original EEC

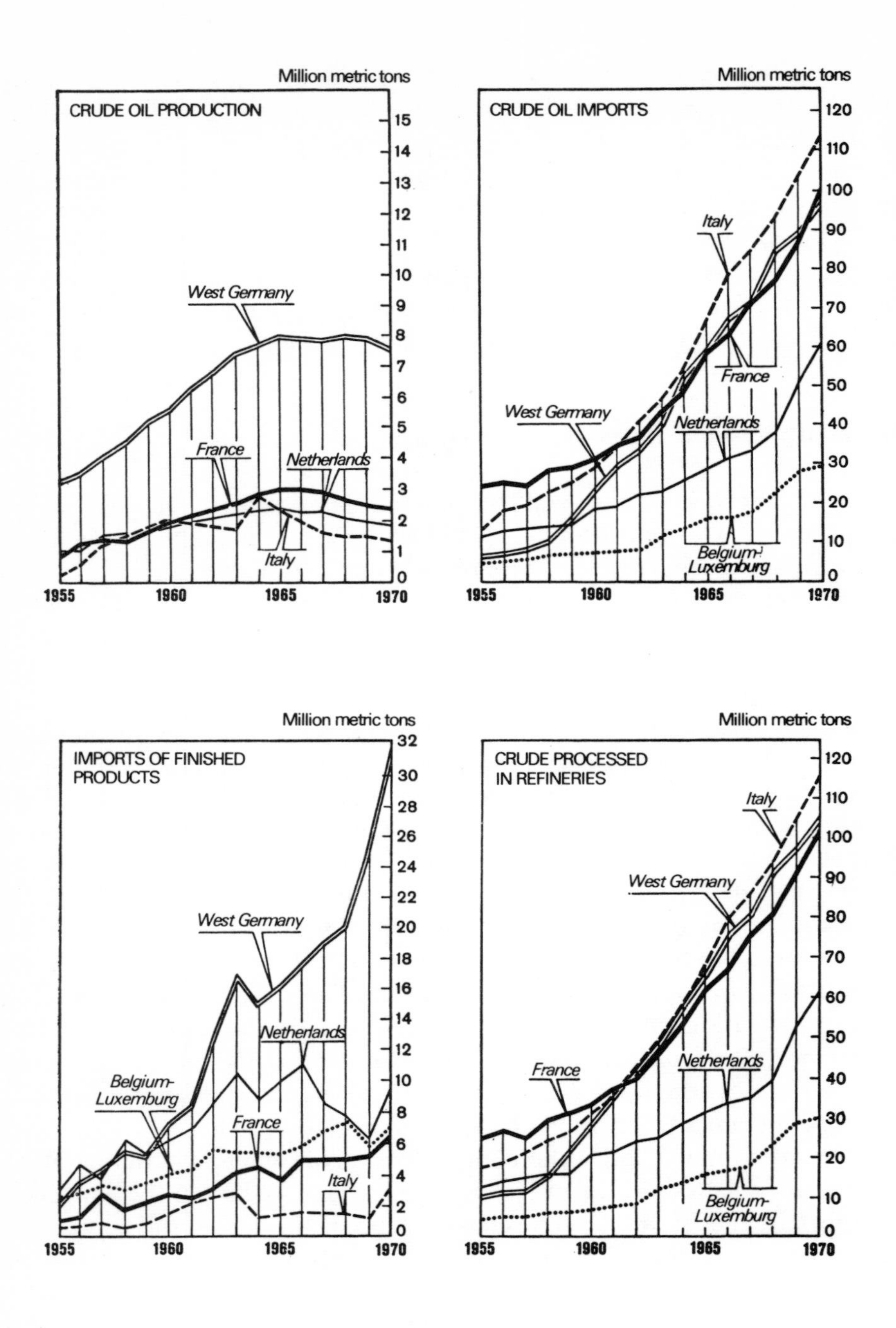

Fig. 20 Oil situation in EEC countries, 1955–70

Source: Comité professionnel du pétrole

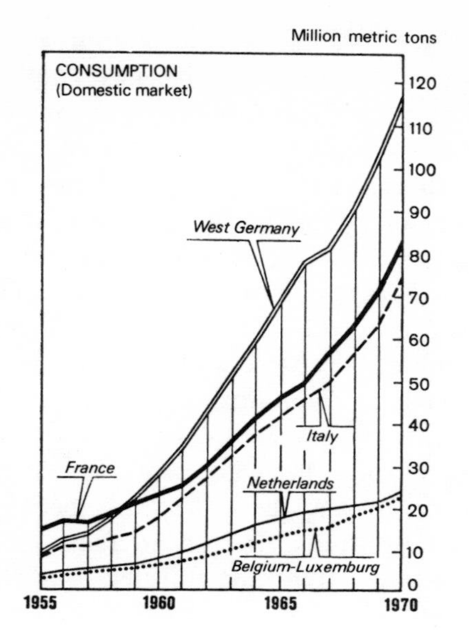
Million metric tons
CONSUMPTION
(Domestic market)
West Germany
Italy
France
Netherlands
Belgium-Luxemburg
120
110
100
90
80
70
60
50
40
30
20
10
0
1955
1960
1965
1970

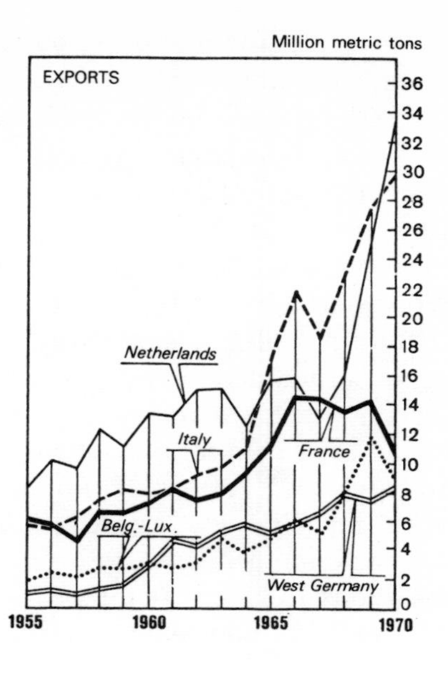
Million metric tons
EXPORTS
Netherlands
Italy
France
Belg.-Lux.
West Germany
36
34
32
30
28
26
24
22
20
18
16
14
12
10
8
6
4
2
0
1955
1960
1965
1970

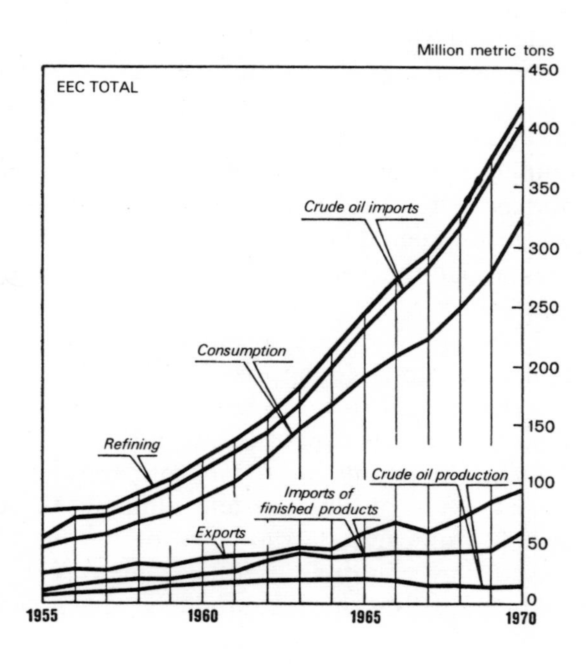
Million metric tons
EEC TOTAL
Crude oil imports
Consumption
Refining
Crude oil production
Imports of
finished products
Exports
450
400
350
300
250
200
150
100
50
0
1955
1960
1965
1970

countries have shown a rising curve since 1955 (the year chosen as the initial one for the period to be reviewed in this book). They totalled 61,777,000 metric tons for 1955 and rose to 402,160,000 metric tons for 1970.[10] This spectacular growth was assisted by the decline of coal, although the governments have tried to prevent the collapse of the coal industry, and also by the lowering of selling prices by the widespread practice of granting discounts. From 1959 to 1969 oil prices before taxation were lowered by an average of between 25 and 30 per cent. The sharp upturn which occurred in 1970 caused the consumer governments some concern but it was not catastrophic, because countries have made efforts since the war to minimise the cost of their supplies, above all by setting up their own refining facilities.

The quantities of crude imported before the war were still small. Refining capacity was created, but not for any great volume, because Europe was importing mainly finished products. It was, in fact, more advantageous to refine at the source and then transport finished products to the European consumer markets. National policies were then taking a course contrary to the usual practice and the beneficial results of these efforts were not apparent until much later.

After the war the problem became much more acute. Official policy had to try in any way possible to reduce the increasingly heavy burden of the bill for energy supplies. Payment had to be made for the most part in dollars and that currency was desperately scarce in devastated Europe. Governments then used the carrot and the stick in seeking to get the foreign companies supplying oil to finance the construction of local refineries and the development of transport facilities subject to control by the purchasing country. Bank loans, permits for the imports of indispensable equipments, preferential tariffs or quota guarantees and assignment of market shares were granted only to companies which agreed to give in to these government demands.

The companies usually accepted these policies, especially as the abundance of new sources of production and keener competition, the decline in the use of coal and great technological advance in transport methods enabled the total quantities imported to be increased considerably. This augmented flow of oil into the countries had the effect of rendering viable the construction of coastal refineries at lower cost and it even became financially acceptable to site refineries far inland, because of the development of a pipeline system running through the large industrial centres.[11] Moreover, the interests of the big companies and those of the importing countries were in line, as a result of the intensification of nationalism in the producer states and the chronic instability occuring in the world's leading oil-producing region. In 1951 Dr Mossadegh had nationalised Anglo-Iranian and the Abadan refinery was closed for more than three years thereafter.

The most prominent feature from 1950 onwards was this increase in refining in European territory, because until then there had been great dependence on products refined in the Antilles, the Persian Gulf or the United States. Refining facilities had been almost non-existent immediately after the war in the six countries which initially formed the EEC but in 1970 there was a capacity of 484·3 million metric tons. Refining also plays a part in overall trade, because sizeable re-exports are often possible: thus it was calculated that for 1969 Italy recovered 40 per cent of the costs of its purchases of energy by encouraging large-scale export from refineries in its territory. In the case of France the volume of re-exports was also large, at some 20 per cent of the value of imports. [12] In that year Spain processed 28·6 million metric tons of crude in refineries in its territory and since the domestic market was not yet absorbing that quantity, some 5 million metric tons of finished products were available for export, whilst imports of finished products were about 500,000 metric tons.

The success of this policy of achieving local development depends on the State's ability to exert telling pressure on foreign companies operating in its territory.

France has undoubtedly been the scene of the fiercest skirmishes with the international groups when the issue was not only to protect the positions of the state enterprises in the domestic market, in the manner of other countries, but also to increase the percentage. Historically, this policy arose because of the discoveries in the Sahara. The free market Algerian crude had to be given a place in the French distribution network, through the setting up of UGP. At the same time the international companies were required to take the Saharan crude as a 'national duty'. This stipulation will not apply in future because of the cessation of supplies from Algeria enjoying special favour (8 per cent in 1973 compared with 26 per cent of the total in 1970). However, the ending of the scheme will not prevent the public enterprise groups from retaining the 50 per cent share of the national market which they managed to obtain as a result of government pressure.

From time to time there are outcries against national *dirigisme* as compared with the easygoing liberalism which has occurred elsewhere. But the international companies have no great cause for complaint about the level of their operations in France. It seems unreasonable to suppose that a national public enterprise which has discovered oil abroad should have some kind of prohibition imposed on it in connection with placing the oil on its domestic market. Such a situation would appear to be particularly absurd under the Common Market policies, which require competitive marketing as a 'national duty' as the state companies become strong enough to progress to international competition. The obligation to take any unsold quantities which CFP was unable to place was stipulated a few years after the formation of that company but already at that time it was not really justified because the company had achieved worldwide operations. CFP developed trade channels enabling it to sell its crude and finished products made by itself and other producers to countries other than France.

Whilst the relaxation of some stipulations deriving from the earlier days of state enterprises might be justified, the policy of *dirigisme* has the merit of some logic. It is a very suitable approach in the development of countries which wish to retain the presence of non-nationals whilst favouring the creation and growth of national groups. Unlike Algeria's attitude, *dirigisme* does not seek to eliminate foreign interests but to cause the encounter between them and national enterprises to yield beneficial emulation.

Proof of the rightness of this policy by taking for comparison the consequences of acting too late can be obtained from events in Germany and Japan. In West Germany the Government became worried about the buying up by multinational groups of the various German interests in the business and tried to build a firm enough line of resistance. The stake of the independent companies belonging to German interests is now only 25 per cent in refining and in distribution. The last survivors of a contest which was unequal for lack of initial state support banded together in a joint company called Deminex and decided to strive for integrated operations. They launched straight away into exploration, acquisition of wells and the purchase of shares in existing concessions according to a method initiated by Spain some ten years earlier. They planned to spend as much as DM 25 thousand million in the period 1971–80.[14]

Japan, another country with very large consumption and scarcely any production, was obliged to revise its traditional *laissez-faire* policy. In 1950 the matter had not yet become urgent, as the share of oil in the nation's energy consumption was only 7 per cent. In 1957 oil consumption was about 15 million metric tons. In 1969 it reached 150 million metric tons and by 1974 it had passed the 200 million metric tons mark. A demand of 800 million metric tons has been forecast for 1985.[15] Since the 1962 Act, the Government has been concentrating on giving Japanese companies a place in the domestic market. In order to obtain 'Japanese oil' a public corporation was set up in 1967. Its tasks are to finance and assist the development of Japanese oil enterprises and their oil output and to take holdings in existing companies already engaged in exploration. The Ministry of Industry and Commerce for its part is trying to promote the involvement of existing private companies or bring about the formation of new ones, with a view to private enterprise entering into the oil adventure.

But unlike the countries of Western Europe, the Japanese Government has been overtaken by the speed of events and has scarcely been concerned with the problem long enough to show foresight as regards timing and geographical region. Of the supplies being received, 90 per cent come from the Middle East and reserve stocks would not enable the country to hold out for more than 17 days (or 45 days counting oil in stock abroad or on board tankers). Having failed to discover 'Japanese oil' yet – and it had been hoped that this would supply 30 per cent of the country's needs in 1975 – Japan is especially vulnerable to change of circumstances to the advantage of the producer countries, as the crises of 1971 and, even more so, of autumn 1973 have demonstrated. After the latter experience, which made the country only too well aware of its dependence on the majors and on the Arab producer countries, Japan set about drastically revising its oil policy. On 6 June 1974 the Petroleum Commission of the Japanese Ministry of Commerce and Industry approved a plan for complete reorganisation of Japan's oil policy. This plan aims at making direct agreements with producing countries to eliminate too great a dependence on the majors. The plan also provides for government control of prices of petroleum products, greater concentration of the refining and distribution companies and the obligatory formation of stocks equivalent to sixty days' consumption.

The Petroleum Commission also proposed the setting up of a public organisation to carry out the direct import of crude oil.

The examples mentioned indicate that even where the State does not participate directly in oil affairs its more subtle indirect action is of fundamental importance. It regulates the activities of the companies by means of a whole set of measures, including the control of imports, exports and price levels, the building up of reserve stocks in order to be able to lessen the effects of sudden emergencies, the diversification of sources of supply, and decisions regarding consent to investment proposals of refinery or distribution companies.

In these matters France, Italy and Spain are confronted by the same problems as those of Japan and the EEC Six or Nine, since all of these countries have little production in their territory and heavy consumption. France and Spain are undoubtedly more *dirigisme* than the other countries in oil matters. In the case of France this is because it suddenly found itself under obligation to find purchasers for the oil discovered in the Sahara. Spain has a long tradition of interventionism in oil. Spain's industrial expansion began late, giving the country the opportunity to profit from the example of the experiences of its neighbours. Moreover it was well able to apply a strategy of its own when assailed by requests from multinational companies eager to obtain sales outlets. As for Italy, it is not much concerned with increasing the national share of the domestic market because it is not self-sufficient in supplies. The usual policy is mainly directed towards a commercial strategy of developing refining capacity to the utmost, in order to recover through exports some of the sums spent on worldwide purchasing. This leads to a further question of the overall cost of the involvement of the State.

In discussing aims, I have said that it is always difficult to measure the precise cost of an interventionist oil policy. The amounts shown on invoices for oil imports do not provide adequate indices for determining the true figures for the cost of such involvement. Importing countries are not simply purchasing countries: the important figure is the difference between imports and exports and the value of the processing which the commodity has undergone between those two transactions. Italy, for example, re-exported in 1969 in the form of finished products more than 27 million metric tons of the oil which it had purchased as crude. Moreover the crude provided by the USSR has been paid for by supplying steel tubes, textiles and machines. Similarly, the recent contracts for natural gas have involved advantageous arrangements.

It is also difficult to value results as regards the flow of currency inwards and outwards. Investment expenditure and operating costs are usually introduced into the matter. If a foreign company makes investments in France, then currency is brought in and normally reduces the sum in the total bill for oil. Conversely, if a French company makes investments in Canada that means that currency leaves France, but if the activities are well conducted the outflow of currency will one day lead to revenue and receipts coming in the opposite direction.[16]

Whatever the result of these accounting notions, it is actually possible to compare the actual cost in currency of the oil supplies of Spain and France for 1969. These two countries are at the upper and lower ends of the scale, whilst Italy comes at an intermediate point.

Spain was able to carry half its oil imports in vessels of its merchant navy. But despite some advance by taking over intermediate operations, the oil economy showed a deficit of 376·2 million dollars.

For France, the Hydrocarbons Division of the Ministry of Industry in its annual report for 1969 gives the following breakdown for the actual cost in currency of French supplies (in thousand millions of francs):[17]

Value before taxation of products consumed	17
Costs paid in France	10·5
CIF value of crude and imported products	6·5
including: freight paid in francs	1
portion of FOB price paid in francs or in franc zone currency	1·1
Residual cost in currency of imported crude	4·3
Other costs in currency	0·1
Savings from:	
international operations of French oil companies (less investments)	2
export business of French undertakings	0·4
Balance in currency	2

Thus, the net cost in currency of 1969 supplies was 2 thousand million francs, which was equivalent to 363·6 million dollars.

It is of interest that in the French balance sheet for oil for 1969, savings in currency and expenditure in francs were together equivalent to nearly two-thirds of the value of imported products. Moreover the actual cost in currency of France's oil supplies was similar to that of Spain but the quantities handled for France were three times greater. These figures provide definite justification of the sustained efforts of the administrative authorities to minimise the cost of France's oil supplies. The statistics help one to understand why Spain is trying by every possible means to strengthen its own potential at a time when its deficit from oil affairs is assuming alarming proportions.

J.C. Colli considers that the efforts to improve the cost situation were worth making. He calculated from the 1968 statistics that imports of crude petroleum products into the French economy were of the order of 1 thousand million dollars and represented more than 10 per cent of total purchases of goods, without counting the capital which left the country for investments abroad. Taking into consideration income from shipment under the French flag and from patent royalties, also gains from holdings in sources of production, the cost in currency of French oil activities probably totalled about 250 million dollars, [18] Colli's study is based on a rate of exchange between the dinar and the franc which does not apply now. However, his calculations show currency savings of three-quarters of the value of imported petroleum products instead of the two-thirds given by the analysis quoted above. But from 1968 to 1969, the years under consideration in the respective studies, imports of crude and petroleum products rose by more than 11 million metric tons. Hence, despite the efforts made, the currency cost of supplies is likely to become heavier over the years, especially after the increase in costs of supplies since the 1973 arrangements. Two possible ways of trying to curb this ineluctible rise may be envisaged. The part of the FOB price paid in francs or in franc zone currency might be increased. Alternatively, the international trade of French oil companies might be boosted.

The second possibility takes time and can be achieved only if French holdings in oil wells and the capital available permit. Therefore it is necessary to take maximum advantage of the first method beforehand.

Accordingly, since 1945 France has doggedly pursued the first line of action. But so far oil production in Equatorial Africa has not been sufficient to weigh decisively in the balance. High hopes were raised by exploration in the Sahara, since oil in that region was in the franc zone. The recent trend of Franco-Algerian relations discourages undue optimism in that connection.

Under Protocol No. 12 to the Franco-Algerian Agreements it was specified that the francs obtained in payment for Saharan oil were to be freely convertible, which enabled Algeria to exchange those francs for other currencies. Naturally, this measure did not apply to any costs and profits of French oil interests unconnected with Algeria. However, the Algerians kept reducing the proportion of income which the French companies could transfer to France. The margins for which the Algerian authorities permitted transfer became very small but the local market offered little opportunity for the useful employment of funds. But Algeria took full advantage of facilities for exchanging francs for other currencies, not least when confidence in the franc was low, prior to its devaluation in 1969. The dinar did not stay in line with the franc when rates of exchange were adjusted after this devaluation. However, Algerian oil had already been yielding less and less francs in the preceding years. Algeria's actions would not matter too much if Algeria had little foreign trade or if the balance of trade between France and Algeria were equal or in France's favour, but this has not been the case.

During the 1960s it was possible for oil-consuming countries to take judicious measures to assist their balance of payments situation and to reduce the outflow of foreign currency. In 1973 their efforts were obliterated by the leap in oil prices and by the increasing share in income from oil demanded by producer states. The result of much higher costs was heavy deficits for the consumer countries. West Germany will doubtless be able to maintain a satisfactory balance of payments but Italy,

France, Spain, Japan and Britain will have a large trade gap with the producer countries. Chapter 7 contains a more detailed assessment of the crisis, which would have been even worse if efforts to correct the balance of payments had not been made beforehand.

Indirect action by the State in both producer and consumer countries is justified in that it enables the vital interests of a nation to be protected. Constraints are justified when they oblige multinational enterprises to conduct their operations in a manner consistent with the needs of the well-ordered economic development of a country.

The important point is how far indirect pressure to the detriment of the freedom of private interests is justifiable. Below one limit, which moves with the passage of time, there is too much variability in the behaviour of private enterprise and the actions of the various operators can never add up to a situation benefiting the general interests of the people; beyond another limit, which also fluctuates, there is such rigidity that the stimulating effects of competition are inhibited and there is a wastage of funds. All the positive aspects of interventionist oil policy are to be found in situations between those two extremes, the most valuable function of such a policy being its continual striving to achieve a balance between instability and inflexibility.

At the same time, oil affairs in the states with Western Mediterranean coastlines are always at risk from new circumstances and new forces beyond the control of the governments of those states.

Notes

[1] A. Murcier, 'Les deux faces des pétroliers américains', *Le Monde,* 23 December 1969.

[2] The Thirty-First OPEC Conference, held in Caracas from 9 to 12 December 1970 passed Resolution No. 120, which recommended that the Member Governments adopt the following objectives: (a) that the rate of taxation of profits of oil exporting companies be brought to not less than 55 per cent (thus going beyond the fifty-fifty rule; Venezuela, for its part, had already unilaterally set the rate at 60 per cent); (b) that posted prices in all the exporting countries be harmonized, by alignment with the most favoured countries; (c) that a general increase in those posted prices be sought, such as to reflect the favourable trend in the world oil market; (d) that new bases be determined for calculating the 'differentials' applicable to posted prices (quality, gravity, geographical position in

relation to the centres of consumption); (e) that suppression of all the reductions in effect in respect of export prices for oil be obtained as from 1 January 1970.

[3] The parties to the Teheran Agreement were: Producer countries – Abu Dhabi, Iran, Iraq, Kuwait, Saudi Arabia, Qatar; Majors – BP, Gulf, Shell Petroleum and Royal Dutch, Standard Oil (California), Standard Oil (New Jersey), Texaco, CFP; Independents – Marathon, Continental, Nelson, Bunker Hunt, Occidental, Amerada Hess, Atlantic Richfield, Grace Petroleum, Hispanoil, Sohio, Gelsenberg, Petrofina, Ashland, Aminoil, Arabian Oil.

[4] Statistical results of the Teheran and Tripoli negotiations (1 dollar per barrel = 40 francs per ton, approx.):

	Persian Gulf		Libya	
	Before Teheran	After Teheran 18 Feb. 1971	Before Tripoli	After Tripoli 4 April 1971
Posted prices (tax reference prices) (cents per barrel)	1·79	2·17	2·55[(a)]	3·45[(d)]
Rate of tax on companies' profits	50%	55%	55%	55%
Prescribed increases per annum (cents per barrel)	–	+2·50% +5	2	On basis of 3·20 +2·50% +7
Duration of agreement	–	5 years	(b)	5 years
Retrospective tax (cents per barrel)	–	–	(c)	

(a) Until October 1970: 2·25
(b) September 1970 agreements made for five years
(c) September 1970 agreements: backdated to 1965 by 5% surcharge on 50 per cent rate of tax.
(d) Including 0·25 dollar variable geographical premium

Source: *Le Monde,* 4–5 April 1971

[5] 'Oil on the brink', *The Economist,* 6 February 1971, p. 62.

[6] *Les Echos,* 10,763, 1970, supplement, pp. 8–9.

[7] R. Mabro, 'La Libye, un Etat rentier?', *Projet,* 39, November 1969, pp. 1090–102.

[8] Ibid.

[9] Ibid.

[10] In 1970 oil imports included the following: Italy – 113,500,000 metric tons; France – 100,160,000; West Germany – 98,300,000; Spain – 35,495,000. See Comité Professionnel du Pétrole, *Pétrole, 1970.*

[11] P.R. Odell, *Oil and world power,* Peter Hall (ed.), London 1970. Quotation here is from the carefully documented analyses in Chapter 5, 'Oil policies in Western Europe', pp. 95–117.

[12] 'Lessons of Europe', *Petroleum Press Service,* February 1970, p. 43.

[13] Comité Professionnel du Pétrole, *Pétrole 1969,* E 39.

[14] *Petroleum Press Service,* December 1970, pp. 440–1.

[15] P.R. Odell, op.cit., p. 121.

[16] M. Byé, 'Politique française d'importation pétrolière', (French oil import policy), addressing Economic and Social Council, session on 23 July 1958, *Bulletin du Conseil Economique,* 24 July 1958, p. 488.

[17] DICA (Hydrocarbons Division of French Ministry of Industry), Report for 1969.

[18] J.C. Colli, 'La politique énergétique de la France', *Le Bulletin de l'économie et des finances,* 46, March–April 1969, p. 16.

6 The Changing Scene

Firmly interventionist oil policies enable the countries on either side of the Western Mediterranean to enjoy some degree of autonomy in international oil affairs. The countries from which the raw material comes are anxious not to remain merely sellers and tax collectors, the consumers are striving to be something more than mere purchasers. It has already been pointed out that these two most important participants in the oil trade see each other as rivals who need at the same time to exercise a certain amount of understanding. Some degree of harmony in their approach to matters serves as a buffer between these opposing interests, enables compromises to be reached and obtains recognition of the fact that the points which are mutually advantageous to the two parties are more important that the initial hostility. Despite the opposing points of view it is possible to attain a minimum of consensus if both contenders are aware that neither can win decisively over the other. Each side has so far found the kind of 'political coalition' described by Riker to be advantageous. But is there sufficient common ground to enable these two contestants to be joint winners of the stakes in the game, yet at the same time allow each to gain the maximum individual advantage?[1]

The situation in the region under review is undoubtedly still a special one but the chances of survival of this pattern of relationships are uncertain. It is evident that oil affairs are constantly reaching out beyond these narrow geographical limits. Exploration for this raw material, trade in it and its use are on a worldwide scale. When the Suez Canal is reopened to shipping and when a giant pipeline runs from the Red Sea to Alexandria, then obviously some specific reasons for the interests of producers and consumers to join together will no longer apply.

Discussion in this chapter will seek to show that the general situation of instability is modifying the pattern in the oil affairs of the region around the Mediterranean which is under special study in this book. Analysis will be centred on three aspects which seem to indicate most clearly the trends for the next ten years. These considerations are: break-up of geographical solidarity, limitations to nationalism in oil affairs, and competition from other energy sources.

Break-up of geographical solidarity

It is the European Common Market rather than Southern Europe that is dependent on supplies from North Africa, but the position of the producer states involved is not invulnerable.

North African oil is, of course, favourably located for supplying the countries of Western Europe with Mediterranean coastlines.

In 1970 Algeria provided 26·9 per cent of France's imports, 2·5 per cent of Italy's and 8·1 per cent of those of West Germany. Libya had a stronger position in the market, as its oil accounted for 41·4 per cent of West Germany's imports, 30·8 per cent of Italy's and 17·6 per cent of those of France. Whilst Italy was dependent on the Middle East for more than 56 per cent, Germany and France obtained only 34·3 and 44·1 per cent of their supplies, respectively, from that region.[2]

Oil statistics since 1971 show a considerable reduction in the market share of North Africa, a marked advance by Nigeria and the consistent achievement by Middle East oil of more than half the market. Western Europe's supplies were more widely dispersed over the world after the semi-nationalisations in Algeria and the manoeuverings to which foreign companies had been subjected in Libya. Western Europe had become manifestly dependent on producer countries east of the Suez Canal. In 1972 Libya supplied less than 30 per cent of Germany's imports, less than 20 per cent of Italy's and less than 10 per cent of those of France. Algeria had only a marginal share of the Italian market but obtained 11 per cent of the German market. For exports to France there was a spectacular decline compared with the 1960s, as in 1972 Algeria delivered only 10,837,000 metric tons to French refineries out of their total imports of 117,793,000 metric tons. From 27 per cent in 1969 there was a decline to less than 10 per cent in 1972.[3] By contrast, in 1972 France obtained 63 per cent of its supplies from the Middle East (including Iran). There was similar dependence in the case of the nine Common Market countries, because out of a total of 596,000,000 metric tons, 54 per cent came from the Arab countries of the Middle East (57 per cent for France alone) and with the inclusion of Iran the proportion was raised to 63 per cent.

Looking at the matter from the opposite end – from the side of the producer countries, EEC countries were clearly very important customers for these vendors. Until 1971 North Africa undoubtedly held a lead over the Middle East in supplying the highly industrialised European countries. This fact may be demonstrated by quoting a few statistics. In 1968 deliveries to present or future members of the EEC represented nine-tenths of Algeria's output for the year, much the same proportion in the case of Libya, two-thirds for Iraq, one-third for Saudi Arabia, a quarter for Iran and only one-tenth for Venezuela.[4] These percentages reveal the importance of Western Europe's purchasing power. That level of trade would offer a possibility for an agreement between North Africa and the enlarged European Common Market. Certainly during the 1960s North Africa took full advantage of the situation created by the closure of the Suez Canal and the difficulties in obtaining shipping space in vessels of larger tonnages – the building of many more supertankers having been a consequence of the closure.

But Algeria and Libya are not very favourably placed in present circumstances. Plan for the 'Sumed' pipeline from the Red Sea to Alexandria are now going ahead, with construction by a consortium of Italian companies and technical assistance from the American Bechtel corporation. The capacity available in 1976 will be 40 million metric tons a year and may be doubled within the following six months. Apparently the level of profitability will be reached when the flow is in excess of 60 million metric tons. The only disadvantage of this means of transport is that it involves many changes of container along the way. The additional tonnage provided by the reopening of the Suez Canal and by its enlargement is expected to be some 300 million metric tons and to involve 5 to 8 per cent of the world oil fleet. From precise economic calculations it will be possible to know in future which means of transporting oil from the Persian Gulf to the ports of Europe is the most advantageous. Possibilities include the use of supertankers going round the Cape, the use of smaller vessels capable of passing through the Canal, or unloading the crude at the beginning of 'Sumed' and collecting it again at the end of that pipeline. Some shippers are even preparing for a new situation, in view of the ambition of the producer countries to refine their crude locally. There is an inreasing number of orders for tankers of 80,000 to 150,000 tons which are capable of carrying either crude or refined products.[5] In sum, crude from the Gulf states will be arriving in growing quantities at the ports of Western Europe. In the competition between producing regions there is no certainty that the positions gained in the markets of Europe by Algerian and Libyan crudes will be retained. In the final analysis everything will depend on the differences in level between the posted prices for the various producer countries. For crudes of the same gravity and the same quality Algeria and Libya will not be able to obtain any greater advantage of Middle East prices than the allowances for geographical proximity based on the freight costs from the Persian Gulf to Europe, the only point really favouring North Africa is that it has light crudes with low sulphur content that are very popular in all the markets, including the American one, for technical reasons and because they facilitate compliance with increasingly strict anti-pollution regulations.

But when there are unreasonable demands the balance is tipped in favour of the more conservative, or more 'reasonable', régimes of the Middle East. Events in 1970 are illuminating in that respect. In that year the Libyan Government checked the expansion of production for the sake of overall increase in its revenue. This pressure yielded the desired results but at the same time it caused a change in rank order by volume of the world's oil exporting countries. Libyan exports rose by only 6·6 per cent for 1970 compared with 1969. From then on Libya's total has been exceeded by that of Saudi Arabia, which Libya had managed to overtake in 1969 after a particularly rapid rise. As for Algeria, in 1970 it lost its position as second largest producer in Africa and dropped to third place. [6] In 1971 the 'recovery' policies of Algeria and Libya resulted in sharp falls in production (22·8 per cent less for Algeria and 17·1 per cent less for Libya).

There is less community of interests between the producer countries and the consumer countries with Mediterranean coastlines than there would appear to be at first; the limitations in space and in number of possible partners are too great. Important events have helped to bring the two sides together (closure of the Suez Canal, lowering of prices before taxation, decline of coal, European expansion). But change in circumstances may cause a shift in trade relations. In fact, Western Europe as a whole is the natural market for North African crudes. Yet even in that enlarged area the flow of trade cannot be completely isolated from developments. Libyan and Algerian crudes have been obtaining the full advantage of their geographical proximity to the highly industrialised countries but within the present decade their favoured position may become less assured. There may be further discoveries of oil in territory under the sovereignty of the purchasing countries (in the North Sea, Bay of Biscay, offshore in the Mediterranean, Greenland, etc.). Offers from the countries of North Africa may be taken up merely for the unsatisfied remainder of an insatiable demand dealing in large quantities. The North African countries will be under constant competition from other producer countries. Some of the competitors will be especially well placed in distance, as Nigeria is; others, for example Iran, Saudi Arabia and Iraq, might be less exacting as regards the level of posted prices and the rate of taxation, because they have immense reserves and can settle for lower revenue per barrel if their total production is continually rising.

Limitations to nationalism in oil affairs

The behaviour of the Western Mediterranean countries in setting up state companies and using them as special instruments has been described. But there are nowadays some limitations to nationalism in oil affairs. The State authorities are at a disadvantage compared with the big international companies; the State companies carry insufficient weight in world competition; for the consumer countries, which are the more vulnerable, the way out of this serious situation can only be found through political and economic reorganisation at European level.

The governments are capable of making plans. The public enterprises are anxious to create a strategy of their own, as may be seen in France, in Italy and in Algeria. Yet state authorities have difficulty in obtaining mastery over an oil phenomenon which evades their control and is still to a considerable degree manipulated by the large international companies. For the time being the consumer countries seem to have the greater problems. In Teheran the producer countries won the prerogative of determining rates of tax and posted prices. Until then the international groups had been fairly successful in retaining control of these lines of strategy. But beyond that victory the producer states find that they are obliged to reckon with the groups; if irritated nationalism leads to a clash with the groups, then the potential resources in the territory of the country concerned may be left undeveloped until better times; if the producer states allow access to the resources, then they surrender part of their freedom of choice, because they are taken into a worldwide context.

Oil is only one important instance of the limited power of state authorities against multinational firms. Nowadays there are signs of widespread weakening of the control and resourcefulness of political authorities in dealing with the new dynamism of economic organisations. The reasons for this can be explained briefly.

Whereas the political designs of the official authorities seem to hamper a national public enterprise, a multinational firm tends to treat those designs as variables; the public enterprises seem to come under constraint but the multinational ones enjoy considerable independence in the choice of alternative strategies in their involvement and in their bases for development, according to the various opportunities available.

The economic power of states over multinational firms remains limited. Each country can intervene to direct the ventures of multinational firms, above all by means of regulations, but an oil company can make states compete with each other and therefore retains a dominant position in relation to them.

Within the area of sovereignty vested in the State a multinational firm retains a measure of freedom. According to economic circumstances, political latitude, opportunities and the international situation there is some scope for such a firm. The requirements imposed by the political authorities of a nation only relate to part of the firm's activity and may be either assimilated or circumvented.

Another of the great problems in oil affairs is the size of investments and the difficulty in raising capital for a large overall sum. A big international company can go beyond the national possibilities and obtain international finance from many and various sources, for example the Eurodollar market.

Where several countries are joined together in an economic community in which one of the benefits is a lowering of Customs barriers between the member states, it becomes difficult to apply any strongly interventionist national policy, because the oil groups may dodge that obstacle by trying their luck with another member state whose leaders seem more sympathetic to the groups. That lowering of Customs barriers means that frontiers become easy to cross and a country which initially resisted the endeavours of an oil group may be assailed not from an outside position but from within an enlarged economic area. This provides the surest means of obliging countries to reconsider attitudes which are too strict. In that way American capital has penetrated into the European Common Market. Gaullist economic nationalism was at first intractable but was forced by necessity to make concessions. However, it is worth noting that paradoxically the Spanish Government is less vulnerable to pressure from the big international groups than its French counterpart. The former is isolated, pursuing a relatively independent policy, and it is able to impose some of its options and manipulate the intense competition between the firms; the latter is linked to a much larger market and cannot risk an independent policy unless all its partners support this.

In the confrontation between producer countries and consumer countries, the state companies on either side have sought to set themselves up as intermediaries of choice, with the ability to take over a rôle which has previously been assumed by the large international groups. The Western Mediterranean has been favoured as a venue for that offensive but it has never really taken hold in Libya and it has not yet acquired the means of spreading throughout the world. Tactical successes and local victories have been achieved but none of the general intentions has triumphed definitely. In order to prevail it would have been necessary to succeed everywhere at once, which has been manifestly impossible. Undoubtedly the supremacy of the seven majors has been challenged outside the United States but not decisively. The ground which they have lost in the 1960s has been taken up by other companies rather than by the state companies.[7] And the other companies (independents, CFP, Japanese enterprises) align themselves with the majors rather than with the state companies. This was clearly apparent in the large-scale confrontation at the beginning of 1971 between the member countries of OPEC and the oil companies.

The state companies of France and of Italy appeared to adopt a peculiar behaviour of their own in that affair. Prior to the Teheran Conference, ENI published an announcement in true Mattei style stating that its operations were guided by concern for the protection of the interests of consumers and that it did not subscribe to the machinations of an international 'oil cartel'. After hesitating for several days, ERAP finally decided, like ENI, not to take part in the Teheran Conference, on the grounds of its special links with Algeria and having regard to the fact that the Franco-Algerian negotiations were taking place simultaneously.

Yet at the beginning of 1971 all observers had emphasised that the future of oil relations and the part which each participant would play was to be considered at the meeting in Iran between the producers, united in OPEC, and the large companies. Neither France, Italy, or Algeria, and neither Sonatrach, Linoco, ELF, nor ENI could afford to overlook such a monumental fact; they have only limited room for manoeuvre within guidelines laid down by the important producer countries and the powerful private intermediaries.

Certainly at the Teheran Conference the majors were shorn of some of their former prestige and pride. Because of the adroitness of the producer countries assembled by OPEC and in consequence of a world situation favouring the holders of sovereignty over the oilfields, the majors were forced to surrender their traditional privilege of deciding posted prices. They were obliged to accept a large rise in the rate of tax on profits, without being able to isolate the Persian Gulf and North Africa from each other. The bargaining power of the producer states was at its zenith during the first half of the 1970s; it will not necessarily remain as strong in future. Because of their flexibility and their adaptability the majors are still indispensable intermediaries; it could even be said that if they did not exist they would have to be invented. The problem for the future is that whereas national governments are at severe disadvantage against private firms with a worldwide scope of action, the oil companies, for their part, have difficulty in meeting the rising cost of exploration and oil production. The Chase Manhattan Bank estimated new investments in the oil industry over the period of the 1970s at 225 thousand million dollars, of which 100 thousand million would be spent on exploration and drilling.[8] It is impossible to imagine how investments of that size could be found other than with the participation of the majors. Even so, large sums will have to be borrowed to finance exploration which is becoming more and more costly at all stages. This may partly explain the agreements reached in Teheran. Doubtless the majors were displeased by an upturn in prices; in that connection they have a tacit understanding with the producer countries. Whatever the truth on that point, there has been a marked improvement in profits since 1971.

The party which emerges worst from this dubious enterprise is the consumer state. In its difficulties it brings into action its base of power in the industrial world, the state company, as soon as the decisions of the giant oil companies yield to the demands of producer countries and challenge its own designs.

The sole means of ending the situation is to make the fields of political action and of economic decisions overlap again. But this shift requires a clear decision on the part of the consumer countries west of Suez to form a Europe having contractual bonds with North Africa and the Persian Gulf and possessing autonomy of action in the economic field and at political level. The voice not heard during the 1971 crisis and the interested party conspicuous by its absence from Teheran and from Tripoli was Western Europe as a single consumer zone.

It is no longer sufficient to concentrate on what is happening in the Western Mediterranean area. The future course of relations between producers and consumers will depend on arrangements at Common Market level.[9]

In an article which attracted considerable attention, three economists with special knowledge of European problems attempted to analyse the profitability of the 'enterprises in the Community' and to compare it with that of the large international groups from third countries, in order to measure the long-term chances of survival of the former in world competition. By 'enterprises in the Community' the authors meant those with headquarters in one of the six Member States. Nine enterprises fell into that category: the Royal Dutch Shell group, CFP, ERAP, ENI, the Belgian group Petrofina, SNPA, ANTAR and two German groups – GBAG (Gelsenkirchener Bergwerke) and Wintershall. Within this group in which only CFP, ERAP, Petrofina and Shell were integrated enterprises, Royal Dutch Shell stood apart, because in 1966 more than 190 million metric tons were processed in its refineries and it held second place in the world rankings of oil companies. With 18 per cent of the Community's refining capacity, the assets of this group were divided between a Dutch holding company with a 60 per cent stake and a British holding company, Shell Transport, owning 40 per cent.

There was disparity between the groups from third countries in their operating results but in overall refining capacity the British and American enterprises improved their share of the European market from 32·4 per cent in 1958 to 40·8 per cent in 1967, whereas the integrated enterprises of the six Common Market countries suffered a decline in their share from 56·1 per cent in 1958 to 49·6 per cent in 1967.[10]

The comparison is even more instructive when one turns from the economic to the financial results. With the exception of the Shell group, the financial structure of the enterprises in the community had three shortcomings by comparison with the means for American groups, namely a marked shortage of funds of their own, a relative deficiency of permanent capital and very low reserves.[11] Moreover, the composition of the permanent capital (funds of their own plus long- and medium-term debts plus provisions) showed for 1965 a high percentage of long and medium term indebtedness. This practice had enabled growth to be stimulated but only at the risk of an increasing indebtedness that would be difficult to bring under control. Only Shell and CFP had adopted a financial policy similar to that of the American companies, in that they financed the greater part of their investments from their own funds, which gave the disadvantage of a decline in their relative share of the market.

The rate of profit of an enterprise shows the ratio of profits to capital invested over a period of several financial years. Only general indications were available in that connection but the authors were able to ascertain that, unlike the international groups, the 'enterprises in the Community' are not in a position to obtain from their own funds the capital which they would need to obtain normal expansion. They are particularly badly placed for the purpose of satisfactorily carrying out increases in capital or issues of stock on the market'.[12]

The enterprises in the Community therefore faced a difficult choice. Most of them were small, unlike their foreign competitors. They should be able to grow if they made savings by operating on a larger scale, because the difference in distribution costs between the large enterprises and the small or medium-sized ones was estimated at one dollar per metric ton.[13] Their inherent weakness lay in the fact that if they wished to maintain or surpass the rate of growth on the European market they must run dangerously into debt. If, on the other hand, they wished to finance their investments from their own funds, this could only be done to the detriment of the market share which they already held.[14]

The basic conclusions of the review quoted remain valid even though the details of the scene have changed since 1967, in particular by ERAP's purchase, with the assistance of CFP, of ANTAR. But the general trends indicated have remained. They are likely to be aggravated by the rise in costs and limitations on production in Algeria for ERAP and CFP. Moreover, Britain's entry into the Common Market will drastically change circumstances. Shell and BP will be brought fully into the Community. On paper, therefore, the Community will have a much stronger oil position than at present. A great deal depends on American reaction to this redistribution of forces. For the time being it is difficult to see how the small enterprises in the Community will overcome their fears on seeing the two British majors enter this enormous market in strength and reinforce the representation of powerful interests. Will a European oil community remain or does the arrival of the British interests mark the beginning of an Atlantic community?[15]

There are two possible policies which the Community could adopt to avert this dangerous situation.

The first policy would involve reorganisation of the individual oil enterprises in order to enable their activities to be adjusted and their expansion assisted. The aim in such case would be to allow these few scaled-up concerns to participate not only in refining and in distribution but also in exploration, production and transport. They would then not be enterprises of the Community in the sense to which the article quoted above referred but Community enterprises, as defined by the Council of the Communities on 10 July 1967:

> The term 'Community enterprises' means enterprises whose basic interests are by nature permanently in accord with those of the Community and which are unable to benefit from the privileges granted in the country of origin of an enterprise from a third country to its affiliates established within the Community. Such accord of interests may be considered to exist where an enterprise is controlled by nationals or a government of a Member State and the centre of gravity of that enterprise is located in one of the countries of the Community.

A European policy would therefore involve promoting the creation and expansion of one or two Community enterprises by so arranging market shares that from the beginning there would be some equalisation of the condition of competition. This would amount to extending to European level the ideas and practices which have obtained in France for more than forty years.

The second possible policy would be akin to indirect state intervention.

This would consist of regulating competition in the European market, specifying the manner in which it was permitted to operate and guarding against concentration capable of giving groups excessive influence.

As this book has already shown, this policy giving the authorities a function of arbitration rather than of direction of affairs is more discreet but just as traditional as direct intervention in oil matters. A basis for such an approach may be found in certain Articles of the Treaty of Rome, especially in Article 86, which envisages the exercise of some surveillance over the functioning of competition between undertakings.

A Joint Declaration dated 18 December 1968 laid down initial guidelines for a Community energy policy. These contain elements of both the policies just mentioned, without really opting for either of them. They emphasise the necessity of assisting the development of Community enterprises by means of a set of fiscal measures decided on by national governments or by the Community, with a view to taking charge of genuinely European sources of production. They propose favouring balance and fair competition between European and non-European enterprises. They aim at preparing the way for a European supply plan, by asking companies to notify the Council of their import and investment programmes.[16] Finally, in a Memorandum dated 22 July 1971 the Commission recommended the Council to proceed immediately to provide for the hydrocarbons sector 'the possibility of recourse to a set of rules similar to those included in the Treaty establishing the European Atomic Energy Community under the title *Joint enterprises*'. The organisational arrangements for an enterprise of this kind would enable certain advantages to be given, such as adjustments in taxation, guarantee for investments in foreign territories and assistance with finance. In return the joint enterprise would undertake to give priority to meeting the needs of the Community in the event of difficulties in obtaining supplies. The Commission stated as a reason for urgency in providing arrangements in the matter the growth in demand for crude oil in the Community from 410 million metric tons in 1971 to an anticipated 720 million metric tons in 1980. More than 95 per cent of the oil consumed was imported from third countries and by 1980 that source would supply about 70 per cent of the Community's primary energy needs.

In brief, the European concepts of 1968–71 reflected, with variations, the two facets found in Spanish, Italian and French oil policies – direct management and indirect regulation. This provides further evidence of the fact that there is no other choice open to a federation of states which has to buy most of its oil from outside the area of its common market. However, these rational ideas are as yet nothing more than principles and general ideas. When it comes to putting them into practice it is difficult to imagine that ENI, ELF, Deminex and Petrofina would enter into collaboration for the purpose of presenting a common front against the giants of the oil world. The Commission has, moreover, adopted the decision to allow free movement of petroleum products within the Common Market. The situation is complicated by the position of Britain in oil affairs. As A. Hamilton wrote in the *Financial Times:* 'From all evidence, the United Kingdom is not obliged to defend the position of the (large) companies, but it is inclined to espouse their point of view'. Although the British Government has almost a half-share in BP, that organisation acts as a purely commercial enterprise. In a climate of anti-British feeling in the Middle East that semi-state company deliberately chose to set itself apart from the State in order to be in a better position to protect its interests. That line of approach is comparable to that of CFP but is completely the opposite of that of ERAP in Algeria and even in Libya and in the Middle East.[17]

For the time being, where oil is concerned the EEC assembles differing national interests and is certainly not a true multinational community bringing to bear the power that would be conferred by acting as a single purchaser. The Teheran confrontation in January–February 1971 and the crises of autumn 1973 occurred when an interested party of great importance, the European consumer state, was not ready to intervene. However, there could equally well have been a fiercer tussle if another notable absentee, the Japanese consumer, had been represented. Will another serious emergency in connection with the security of supplies be needed before Europe wakes up to the fact that Clemenceau was not so far wrong when he wrote to President Wilson at the height of a world war that, 'fuel is as necessary as blood in the coming battles'?

Competition from other energy sources

Special relations between the Western Mediterranean coastal states arose from the importance of oil to both sides. After encountering strong resistance at first, oil prevailed over coal. Over a period of twenty years the two fuels have changed places in order of magnitude of use as primary energy sources in Western Europe. Oil has emerged with a clearer lead in France, Italy and Spain than in Britain and West Germany.[18] If the prosperity of a country may be measured by its oil consumption in relation to the number of inhabitants, then France and Italy are high on the world scale, whilst Spain is in an intermediate position, as figure 21 shows.

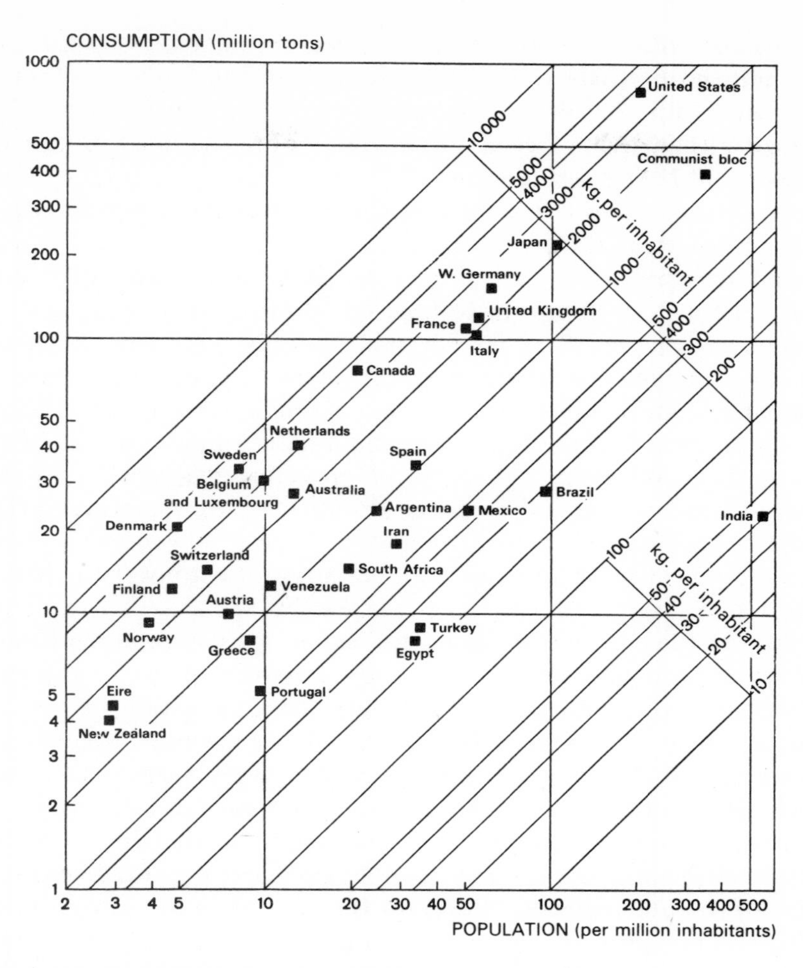

Fig. 21 Per capita oil consumption in the world, 1971

Source: World Petroleum, September 1971

But what would happen if in the future oil lost its present dominance? Its present supremacy may be challenged in the years ahead by considerations of three different kinds which are not merely economic but also psychological and political. Such considerations are basically connected with the inherent characteristics of the various energy sources, the security of supplies required for those sources and the ways of financing them.

The causative factors for the success of oil over coal are above all oil's inherent characteristics. It is a 'clean' raw material, easy to handle and to transport and with a very high calorific power.

For all those reasons it is commonly estimated that 'oil remains competitive with coal even if its selling price per calorific unit is 10 per cent higher than that of coal (for thermal power stations) or between 5 and 8 per cent higher for other uses'.[19]

But as rivals of oil, gas and nuclear power are as good, or even sometimes better, in those respects.

Liquid natural gas has advantages by comparison with oil. There is preference for it, even at a price level up to 10–15 per cent higher than that of oil.

As for uranium, one of its merits is saving in storage space and cost. Commenting on the fact that the most economic way of stockpiling energy is to store it in the form of uranium, P. Desprairies has pointed out that, 'A stock of uranium capable of providing the whole of France's electricity consumption (if nuclear electricity) for five years would fit into a medium-sized hangar'.[20] An advantage of nuclear power is that although it discharges hot water into rivers it does not cause atmospheric pollution as oil does.

Following the example of the United States, legislation in the European countries is becoming stricter as regards curbing damage to the natural environment by various means, including the combustion of petroleum products. There has recently been more talk of inventing a new kind of vehicle engine that would not require petrol. The oil business has obtained part of its wealth from the growth of motoring but is it going to be abandoned by vehicle-makers? That may happen although not for some time yet. Meanwhile improvements are being made, in order to reduce the output of noxious fumes. However, the use of petrol only creates a relatively small problem compared with the effects of the growth of industrial employment of fuel. The dangers in that connection are more serious, so that the oil industry is preoccupied with overcoming them and is itself financing research in that field. Undoubtedly selling prices for manufactured goods will in future have to allow for the cost of measures to reduce atmospheric pollution and some of these measures will inevitably raise the price of industrial fuel.

The objective characteristics of a product are not sufficient by themselves to cause its preeminence. The product must also be easily obtainable. Europe imports 95 per cent of its oil and more than 60 per cent of the proved reserves in the world are in the Middle East, which is a region of relative political instability.

It has been thought that the American and European discoveries in Alaska and in the North Sea could serve as sufficient answer to the mounting claims of producer countries in other parts of the world. However, for the moment such ideas are illusory. It has been calculated that the deposits in Alaska would supply about 100 million tons in 1975 but this additional energy will not even satisfy increased demand on the US market.

Similarly, miracles must not be expected from the North Sea. It may reasonably be anticipated that the Ekofisk field will yield 50 million tons in 1974–75. The Forties field, off the east coast of Scotland, will supply 12·5 million tons in 1975. But the total will only be equivalent to 6 per cent of estimated needs in Western Europe for that period. Production from the North Sea may be put at 150 to 200 million tons by 1980. Meanwhile European consumption is growing every year, so that an annual equivalent of the output of two fields like Ekofisk would be needed merely to keep supply level with the forecast annual rise in European demand.[21]

Gas has been of greater interest since large quantities have been found in North West Europe, thus reducing the European problem of dependence on extraneous sources of energy. The predicament is also alleviated by the fact that uranium is fairly widely distributed in the world and occurs in countries where oil is not present in the subsoil. France, for example, has access to the deposits in the Aïr uplands of Niger, corresponding to twice the energy reserves at Hassi-Messaoud and equivalent to 10 per cent of known world uranium reserves. Moreover, currency costs for uranium are remarkably low. 'Mining accounts for only 10–15 per cent of the cost per kilowatt: the remainder arises from advanced technology and concrete. A kilowatt of nuclear power is therefore 85–90 per cent national as regards investment, labour input and the cash spent.'[22]

States, like individuals, do not like to depend on others for their essential supplies. Despite the precautions taken by the consumer countries (diversification of sources of supply, setting up public enterprises, stockpiling) it is impossible to obtain completely assured supplies of imported oil. Indubitably consideration of the security factor explains delay in the demise of coal production in Europe and the heavy investment in nuclear research and facilities. Energy obtained from national resources is generally preferred to imported energy.

The decisive factor is the size of the gap in selling prices. Expert assessments vary on the matter of forecasting when nuclear energy will be competitive but some experts envisage that from 1980 onwards large-scale production will permit the supply of nuclear energy capable of competing in price with oil-based energy.

One of the unknown factors in connection with such predictions is how oil prices will move.

The price of oil

During the 1960s the reduction of oil prices, even reaching a drop of as much as 30 per cent before taxation, helped to confirm the predominance of oil. The rises since 1969 should in time benefit gas carried by pipeline and even liquid natural gas, the cost of which is slightly higher, but which is being consistently reduced by technical progress and by increasing the quantities processed. Beyond a certain threshold of about 13 dollars per metric ton gas also competes with nuclear energy for electricity production. 'Whereas liquid natural gas is too expensive when energy prices in Europe are low, it thus comes up against a new rival, nuclear power, when the prices of "classic" forms of energy reach a certain level'.[23]

At present there is every indication that we are now in a phase of rising prices. There are technical reasons for that rise: drilling is now reaching greater depths; even if the yeild from existing discoveries is tending to improve, exploration has moved to more remote regions, for example Alaska; moreover the search is increasingly conducted in offshore areas, which are already producing more than 20 per cent of the crude extracted in the world. It is forecast that by 1980 more than 30 per cent of total world oil output will be from underwater resources. Already the majors are making arrangements accordingly: often more than 50 per cent of the area of their exploration licences is offshore. In 1974 a group like ELF/Aquitaine will spend more than 70 per cent of its exploration and production investments on offshore operations, making a sum of more than 2 thousand million francs. But costs are soaring. 'On land, drilling for oil costs 10 dollars per ton of production in Saudi Arabia and may reach 30 dollars in countries with less rich deposits; in the North Sea the cost rises to 75 dollars. For shale oil 150 dollars a year per ton of oil produced is mentioned, whilst the investment for oil from deepwater deposits has not yet been determined clearly but is thought to be of the order of 150 to 200 dollars'.[24]

However, the reasons for rising costs are above all political: the producer countries expect to impose the maximum levy on operations performed in areas under their sovereignty. Including the increased proportion of 'shared oil', which is more costly than oil from the traditional concessions, the part of the price attributable to deliberate policy of the producer states has since 1971 risen in a steeper curve than that for technological change and move to different exploration areas.

There are several explanations for the claims of the producer countries. Governments in command of such a sought-after raw material definitely intend to take advantage of the comparative shortage of available sources by extorting benefits from the companies. They are probably aware at the same time that the age of supremacy of oil will not last much longer, so they seek to strike while the iron is hot. But the companies find that the dramatic situation merely enforces some necessity to humour the host countries. Recent cost increases will be recovered from their selling prices. The operators most affected by the developments are the European state companies, which will find their room for manoeuvre dangerously restricted. The claims for justice from the most militant producer countries have for the time being had the paradoxical effect of actually enhancing the power of the giants of the oil world, whose opportunities for gain are very much greater than those of other oil companies.

If prices rise too steeply over the next ten years, then the big international companies will extend their field of activities to include uranium. ENI has already set up a nuclear energy branch of AGIP for finding and processing uranium but its means are not on the same scale as those of Gulf Oil, Shell and Standard Oil of New Jersey, which are now closely concerned with possibilities in the sphere of nuclear energy.[25]

The real losers from soaring prices might be the major importing countries. Either they may resign themselves to limiting their revenue from finished products, in which case they will have to find other means of indirect taxation, or they may pass on rises, as they occur, to consumers. Because of the isolation of the American market this will place heavier burdens on industry in Japan and Europe than in the United States, whilst the funds of a few large groups, most of them with American headquarters, will prosper.

There is a risk of the ever-increasing amounts required for oil investments drying up the capital market.

It will be difficult to decide which type of energy to support in future. Pierre Guillaumat has justifiably pointed out that 'competition between oil and the atom is still limited in the matter of supplying energy but will be enhanced at the level of seeking funds'.[26] It will be necessary to choose between fairly low-priced nuclear fuel and expensive petroleum fuel. The relative costs are in the reverse proportion in the matter of research. The sums spent on scientific research at present are much higher for nuclear power than for petroleum-based energy. Governments will soon have the difficult task of deciding priorities in research programmes and of assigning funds according to energy plans, for which careful consideration will be given to probable cost trends for the various sources of energy.

Although long-term prospects are less encouraging, as a dynamic worldwide business oil will still enjoy some prosperous times yet. The future depends on the result of interaction between OPEC, the companies and the consumer countries. In Teheran the first signs of crumbling of the previous edifice were apparent. The North African countries will not be able to exert greater influence unless the Persian Gulf states let them go their own way. If they do, then they will have to choose plans that are likely to be tolerated by the European consumers nearby. As D. Bauchard has emphasised, it is an unwritten law of oil affairs that if a source of production is made too costly or is nationalised, then in due course another source elsewhere benefits. With deliberate exaggeration he has written that 'Kuwait was an "invention" of oil operators after the Mossadegh affair; Libya was an "invention" arising from the closure of the Suez Canal'.[27] If Algeria adopts an intransigent attitude on taxation and if Libya proves to be too greedy, then attention may be transferred, perhaps to Nigeria.

A complete change may be caused in another way. The Europeans might perfect their strategy and finally engage in concerted action to protect their common interests. They hold only one card but it is the highest trump. They could band together as a purchasing unit which has means of payment and is very particular about the goods offered, having the power to open or close access to a larger market than that of the United States or of the USSR. This enormous buying organisation should be able to find ways of provoking competition between suppliers and of providing worthwhile rewards for those who offer the best conditions.

Since oil became a strategic raw material vital to the economy and the defence of a country, no government has been able to disregard its importance. Over the years two attitudes have arisen. The State has sought to be an arbiter, dominating the interplay of economic interests and imposing a kind of code of conduct. Later the State itself entered into economic competition, becoming an entrepreneur participating directly in the management of oil affairs. It has been possible to carry on these two forms of activity simultaneously so long as the political function of the State controlled the economic sphere. But nowadays in the hydrocarbons sector relations are more widely spread thoughout the world and geographical proximity is losing its strategic importance.

The struggle between the contenders for the benefits to be gained from oil is becoming more determined. Producer countries and consumer countries are competing with each other more strongly for shares of the profits from oil operations. The international groups are above all afraid of agreement to their detriment between the holders of the sources of raw material. But it seems that at a time when their intervention could make a great difference Governments have lost their flexibility and contracted some kind of inertia. In the face of firms operating on a worldwide scale and having purely economic interests, governments and state companies should join together in a way which enables them to find a suitable answer to this new kind of challenge presented to politics by economics.

Notes

[1] W.H. Riker, *The theory of political coalitions,* Yale University Press, New Haven 1962.

[2] Comité Professionnel du Pétrole, *Pétrole 1970,* E. 45.

[3] Comité Professionnel du Pétrole, *Pétrole 1973,* p. 373.

[4] *Petroleum Press Service,* March 1970, p. 83.

[5] *Le Monde,* 19 January 1974, p. 4.

[6] *Crude oil production* (thousand million metric tons)

Year	1968	1969	1970	1971	1972	1973
Iran	141,791	168,235	190,000	227,346	252,339	293,800
Saudi Arabia	140,998	148,839	175,500	223,515	285,580	364,685
Kuwait	122,085	129,549	138,000	146,787	151,098	138,255
Libya	124,524	149,084	159,000	132,250	107,751	104,482
Algeria	42,145	43,824	46,400	36,346	50,085	51,000
Nigeria	7,298	26,627	53,000	75,306	89,784	100,917

[7] Shares of markets other than US and USSR. (Source: *Le Monde,* 11 November 1969)

	Date	Production	Refining	Distribution
The seven majors	1963	68·2	64·3	62·6
	1968	65·5	59·4	57·5
Other companies	1963	21·9	22·7	26
	1968	24·3	25·6	29·3
State companies	1963	9·9	13	11·4
	1968	10·2	15	13·2

[8] D. Bauchard, *Le jeu mondial des petroliers,* op.cit., p. 37. These estimates excluded the Communist bloc. In its 1970 study, Chase

Manhattan Bank stated that investments for the 1970s should be double those for the 1960s. It also pointed out US preponderance in this matter, because those companies contributed 77 per cent of funds for exploration for new oil deposits.

[9] M. Albert, G. Ascari and G. Brondel, 'La faiblesse des structures financières des compagnies pétrolières européennes', *Direction, 149,* April 1968, pp. 388 et seq.

[10] Ibid., p. 392.

[11] Ibid., p. 395.

[12] Ibid., p. 397.

[13] Ibid., p. 434.

[14] Ibid., p. 437.

[15] C. Tugendhat, *Oil, The Biggest Business,* Eyre & Spottiswood, London 1968, p. 291.

[16] 'Une première orientation de la politique énergétique de la communauté', *Revue française de l'énergie,* 208, February 1969, pp. 219 et seq.

[17] A. Hamilton, 'La politique communautaire suscite en Angleterre plus de scepticisme que d'opposition', *Financial Times,* translation in *Le Figaro,* 31 October–1 November, 1970, pp. 11 and 12.

[18] Comité Professionnel du Pétrole, *Pétrole 1970,* E 5. Consumption of energy, figures for coal and for oil for 1969.

	W. Germany	Britain	France	Italy	Spain
Coal	47·8	54·5	30·0	9·7	24·8
Oil	44·1	38·4	51·8	64·9	52·6

[19] D. Bauchard, op.cit., p. 13.

[20] P. Desprairies, 'La concurrence de l'uranium et du pétrole', *Revue française de l'énergie,* 223, July–August 1970, p. 467.

[21] 'Too little up North', *The Economist,* 20 February 1971, p. 69.

[22] P. Desprairies, op.cit., p. 467.

[23] 'Le GNL ou gaz naturel liquéfié', *Bulletin Mensuel d'Informations ELF,* no. 11, 25 November 1970, p. 8.

[24] F. Didier, 'La recherche et l'exploitation des hydrocarbures en mer. Quelques problèmes spécifiques', *Bulletin Mensuel d'Informations ELF,* no. 5, Supplement, 25 May 1974, p. 5.

[25] 'The oil companies enter the nuclear field', *Petroleum Press Service,* October 1970, p. 375.

[26] P. Guillaumat, op.cit., p. 1429.

[27] D. Bauchard, op.cit., p. 31.

7 The Revolution of October 1973

Since the beginning of the present decade the main focus of oil problems has shifted from North Africa to the Near East and the Middle East. The Yom Kippur war was the occasion when the Arab countries chose to bring oil into the modernised arsenal of political weapons. But the true causes of stiffer demands from producer countries seeking to achieve final breakaway from the colonial past included monetary troubles and world inflation. As often happens when policy is relaxed so that there is a drift into straightforward day to day management of affairs, sudden attack took by surprise and routed the main victims. Until then serious consideration of the growth of the Third World and of trade with it had been confined to academic discussion and took place in specialist circles. Now it has entered the public arena. When Sicco Mansholt expounded his views several months earlier there had been general incredulity. Now some of his predictions have come true in a manner which has shocked conventional policy-makers. No policy can claim to be effective if it proves to be incapable of allowing for the hazards of fate and constantly providing for them in its ideas and decisions.

A political weapon

Indignation at the use of oil as a weapon betrays naïvety, incompetence or forgetfulness. At the beginning of this book it is emphasised that since oil was first discovered it has been a raw material around which relations of co-operation and of conflict between the men and the private and public undertakings involved have formed. When Western governments or large companies controlled the levers of power the weapon already existed but was in the hands of the industrialised world. Everything has changed now that the Third World has realised its strength and is organising itself to combat a form of exploitation to which it is subjected. Signs of disturbance of a world order which had become unacceptable have not been lacking since the end of the Second World War. Dr Mossadegh's experiment in Iran in 1951 has surely not been forgotten so quickly. That attempt was a fiasco but the intention was to expropriate foreign capital and nationalise the resources in the national subsoil. Near home, Algeria's insistence on naturalising hydrocarbons which were providing a profitable business for foreigners should have prompted the oil world to amend its anachronistic attitudes.

In connection with the first two Arab-Israeli wars, in 1956 and 1967, the threat of using oil as a weapon was made for a few weeks but this proved to be only a warning. The United States was able to assist with supplies for Europe and the flow was restored. Despite the tough negotiations in Teheran and in Tripoli at the beginning of 1971, the Western side subsequently regained a certain optimism: there were few who took warnings from the Arab world seriously, although these were as plainly stated as the threat which appeared many months before the event, on 5 November 1972, in a Kuwait newspaper, which declared that 'oil is our ultimate and most powerful weapon'. For that weapon to become effective it needed the opportunity provided by a further war with Israel, solidarity between Arab countries with fairly disparate social and political structures and a sellers' market.

Within a few months effectiveness of oil as a weapon had been demonstrated by the Arabs, because of their unity in collective decisions, their shrewd planning, the very good knowledge of the laws of the market acquired by their leading experts at Anglo-Saxon or French universities and their skill in applying economic manipulation and political pressure. The political weapon has become dangerous because it can be used without open argument, because it reveals cruelly the weaknesses of advanced industrial economies that are too dependent on foreign sources of energy supplies and because it is employed in a clearly demarcated area in which there is more interdependence than ever amid the fiercer rivalries between world powers.

Thus in oil affairs the usual signs of the turmoil of war are now apparent. The process is fully under way. In order to allay the anxiety of public opinion, when it has been brutally shaken out of the over-confidence caused by constant growth, scapegoats have had to be found. Hence the reproaches uttered against Gaullist France for having ceded the Sahara in the Evian agreements; hence the loss of composure by Japan, which felt obliged to denounce the misdeeds of the 'Zionists' in order to remain in favour with their mortal enemies; hence, above all, the humiliation of Europeans, who referred to the spectre of Munich, with its bitter memories of successive capitulations to a brutal, arrogant and insatiable aggressor.

All the accused have hastened to plead not guilty and to put the blame on the supposed provoker of this Middle East upset, the United States. But supported by his adviser Mr Kissinger, President Nixon was having none of it. At the same time the Americans warned the Russians not to fan the flames. *Sovietskaya Rossiya* retorted that: 'There is a repetition of what has happened more than once before: some imperialist forces want to make up for their failures and their disappointments by launching an anti-Soviet campaign, resorting to fantasies. In looking for the hand of Moscow where it is not they overlook the fact that the culprits are known; they are the imperialist and Zionist circles which obstinately oppose a just settlement in the Middle East'. The Moscow press played Pontius Pilate, asserting that it was ridiculous to imagine that King Feisal, Colonel al-Qadhafi and the Emir of Kuwait were Soviet agents.[1] China did not want to be left out of an ideological wrangle which might serve to preserve or win support. It came out in full support of the Arab cause: 'Oil is the form of wealth belonging to the Arab countries, it has been a major cause of foreign aggression. At present the Arab countries, united as one in the fight against the common enemy, are using that resource as a weapon playing an important part in the struggle against Israeli Zionism and against its supporters.'[2]

As for the Arabs, they did not choose to regard the gestures of appeasement from Europe at the end of 1973 as anything more than feigned friendship or the agreements signed previously as being more than 'pieces of paper'. Whilst some circlės promoted the idea of the possibility of an oil war waged by an efficiently led invading commando force, President Boumedienne acted as spokesman for the other side, replying that: 'We do not need armies or tanks to defend our oil. A small number of our fellahin would suffice to provoke a world catastrophe if the West should become obstinate. I would even say that if the West tries to act arrogantly or to use force it will suffer a catastrophe. All the wells will be fired: all the pipelines will be destroyed. And the West will pay the price.'

The extent of changes in oil affairs

In this atmosphere of 'Cold War' on the oil front all concerned have endeavoured to press their advantage or minimise their losses by sheltering under the cover of grandiose statements or hiding behind a smokescreen of ideological clamour. It is necessary to seek out the truth, on the assumption that matters have not necessarily been as bad as they have seemed. When the energy problems of the world arise in such conditions, public utterances may mask the fact that compromises are being worked out or that strategic withdrawals are occurring. Oil politics are, after all, only one aspect of contact between the Arabs and the West; they are only a special area of relations between underprivileged countries and rich countries. But it is certain that oil will be all the more readily used as a weapon in future if international politics lacks means of action and if the Israelis and Arabs continue to watch each other with their firearms ready for instant use.

Even if the general situation involves far more than strictly oil matters, oil will certainly remain one of the essential factors in confrontations in the world during the present decade. We should take advantage of a relative lull in the fighting on the hydrocarbons front to assess more exactly the size of the earthquake which has shaken the European and Japanese economies. We should now try to draw up an initial balance sheet and make some projections of future effects of the events which some analysts have already termed 'the October 1973 revolution' in oil affairs.

Some statistics are necessary in order to assess the present situation, because phenomena of domination and of dependence are more meaningful if one has precise measurements. In that connection the figures show a sharp rise in oil consumption over the last 20 years. From 1950 to 1960 it doubled, from 500 million metric tons to 1 thousand million. It was 2 thousand million in 1969 and had reached 2·5 thousand million in 1972. It was anticipated before the 1973 crisis that consumption would be doubled again during the present decade, which would give a total for the 1970s of 30 thousand million metric tons, which is equivalent to the entire world production since the beginning of the oil industry.[3]

According to the experts, before the 1973 crisis normal expansion of the market would have brought oil consumption in the United States to 1,320 million metric tons by about 1985. An increasing proportion would have had to come from imports (the most pessimistic observers mentioned 50 per cent), because domestic resources could not meet such a demand and Alaska was not the Eldorado for oil that had been promised by some advertising campaigns. The United States is therefore seeking new sources of oil supply and large proportions of the holdings of the big American companies are in the Middle East.

Also present in that region are the Japanese and the Europeans, whose dependence on foreign raw materials for energy is vital in this age of predominance of oil. At the end of 1973 the Japanese found that the American companies preferred to supply the United States when their country was under embargo rather than supply Japan, which was a regular customer of those enterprises but for the time being lacked sufficiently effective means of pressure. It was calculated before the 1973 crisis that by 1980 Europe's needs would rise to 1 thousand million metric tons and those of Japan to 465 million metric tons.[4]

At the beginning of 1974, 55 per cent of world proved reserves were in the Middle East. Saudi Arabia alone held more than 20 per cent of the world total. Its production was 285 million metric tons in 1972, amounting to 11 per cent of world production, and there was an increase of more than 36 per cent for the first eight months of 1973. That country had set itself the target of raising its output of black gold to about one thousand million tons in 1980. Saudi Arabia is a member of OAPEC (Organization of Arab Petroleum Exporting Countries), which has its headquarters in Kuwait and sparked off the 'October revolution'. The ten member countries[5] of OAPEC represent 30 per cent of world crude production (and the addition of Iran to the figures for the OAPEC countries brings the total to 40 per cent). Among world proved oil reserves of 90 thousand million metric tons those in the Arab countries were 50 thousand million metric tons and Iran had a further 9 thousand million. Dependence on oil therefore means dependence on the Arab countries. Crude oil exports from the Arab countries alone provided between 6 and 10 per cent of supplies used in the United States, 85 per cent of those to Japan, 65 per cent for Europe and 84 per cent for France.[6]

The above figures show why Saudi Arabia's support for the tougher Arab demands was decisive for their success. The vulnerability of the economies of Japan and the countries of Western Europe in time of tension with the producer countries is also apparent. An embargo delivers a heavy blow to those consumer countries but in the short term it is not serious for the United States.

With demand growing rapidly and supply having difficulty in keeping up, if the producer states merely impose new arrangements and for a time reduce the quantities available at the wells or at the pipeline terminals such action is sufficient to strike the companies in their Achilles heel and cause them to yield to price rises and tax increases which in less difficult circumstances they would have categorically rejected. Libya had already envisaged, and to some extent tested, the efficacy of such methods in 1970, in seeking to oblige the 'independent' company Occidental to accept its conditions. This policy of giving and taking away enabled the producer countries to get their way in the 1971 negotiations and brought about the end of the undivided rule of the large international companies.

Since then the strategy of the Arab countries has been deployed on three fronts, those of taking holdings in foreign companies, of price rises and of curbing production.

The Mossadegh experiment failed in the world Cold War situation but twenty years later Algeria succeeded in operations of the kind. After patiently nibbling away at the foreign companies and then the French companies, the Algerian Government took advantage of a time favourable to producers to obtain mastery over enterprises operating in its territory by taking full control (100 per cent) or a majority holding (51 per cent) of the existing capital. The Algerians had taken their time over the matter in order to familiarise themselves with trading arrangements and oil technology by learning from the French. When the change-over took place the case was not confined to Algeria; it proved to be contagious, because in January 1973 Iran, with different possibilities, limited the consortium's rights which had been the eventual outcome of that very attempt by Dr Mossadegh in 1951. Iran demanded and negotiated the right to limit the power of action of the companies to sign long-term supply contracts for crude to be collected from the Iranian ports. Any further challenge to sovereignty over oil in the national soil and subsoil was debarred for ever. In March 1973 Iraq, in turn, struck heavily by nationalising the Iraq Petroleum Company (IPC), although it temporarily spared that enterprise's affiliate, the Bassorah Petroleum Company. However, at the outbreak of the October war Iraq seized the holdings of the US Companies – Esso International and Mobil, 23·75 per cent – in the Bassorah Petroleum Company (BPC). On 21 October Iraq went on to nationalise the Dutch interests, namely 60 per cent of Royal Dutch Shell's 23·75 per cent holding in BPC, on the grounds of 'punishment inflicted on the Netherlands for their support of Israel'. A few weeks later the Iraq Government decided to nationalise the 5 per cent share held by the Gulbenkian interests in BPC. Iraq's holdings immediately after these moves were 43 per cent and only three other shareholders remained (BP, 23·75 per cent, CFP, 23·75 per cent and Shell, 9·5 per cent).

On 15 June 1973 Kuwait challenged the agreements signed in New York in October 1972. The laborious compromise made then provided for the Gulf States to take a 25 per cent holding in foreign companies in 1973 and for the gradual rise of this proportion to 51 per cent in 1982. But the process was greatly speeded up. In 1974 Qatar was the first producer state in the Persian Gulf to have signed and ratified agreements providing for an immediate 60 per cent holding instead of the 25 per cent initially arranged. Kuwait obtained the same percentage, although some members of the Constituent Assembly demanded that the holding taken should be 100 per cent and that foreign capital should be altogether excluded. Abu

Dhabi has adopted the same general policy but has waited to see the outcome in Saudi Arabia.[7] In the latter country, the American companies Esso International, Standard Oil of California, Texaco and Mobil have assented to the Saudi Arabian State's holding in ARAMCO being increased from 25 per cent to 60 per cent. Complete nationalisation of the oilfields as in Iran, is envisaged but Saudi Arabia would undertake to re-sell to ARAMCO more than 80 per cent of the crude extracted from ARAMCO's former concessions. The process has spread to Nigeria too. In June 1973 Lagos had already acquired 35 per cent of Shell-BP, a partnership which accounted for about 60 per cent of total production in Nigeria (101 million metric tons in 1973). Admittedly, the Nigerian Government had made a move in that direction two years earlier, when it took a one-third holding in the capital of AGIP/Phillips and a 35 per cent holding in SAFRAP, an affiliate of ELF. In 1973 the Nigerian Government confined itself to mentioning increasing its holdings to 51 per cent by 1982, through the state company NNOC (Nigerian National Oil Company). But on 18 May 1974 Nigeria announced that it had reached agreement with the Nigerian affiliates of the five largest oil enterprises operating in its territory – Gulf Oil, Mobil, AGIP, ELF and Shell-BP – to raise its holding in their capital to 55 per cent.

Thus, the process of nationalisation brought into operation in 1973 and 1974 may be described as an accelerated version of the moves made earlier in Algeria. Libya is, of course, at the forefront of that militant action. Colonel al-Qadhafi showed greater extremism than the political leaders of other countries in that connection – a few months before the Revolution of October 1973 he nationalised 51 per cent of the American companies Esso International, Texaco and Mobil and of the 'independents' Occidental and Oasis, the two last-mentioned providing 15 per cent and 30 per cent, respectively, of a national production estimated at 104 million metric tons in 1973. The position in May 1974, following the complete nationalisation of Bunker Hunt and Texaco, and then of Shell, in the same way as that of BP at the end of 1971, gave the Libyan Government a stake of approximately 60 per cent in the overall production. Thus, the 51 per cent level had been decisively exceeded. In the OASIS group, for example, the State then had a 59·2 per cent holding, because Shell, which had been wholly nationalised, had held an 8·2 per cent stake in that company. The percentage reached in May 1974 may be expected to be raised further, in view of events in the Gulf States and in the light of the fact that Libya has always sought to be the pioneer, rather than a follower of the example set by the producer countries of the Persian Gulf.

However, it is not always to the advantage of the Arab countries to apply nationalisation hastily, as they may have a shortage of skilled technicians and of real knowledge of how to manage the complicated arrangements of international trade in crude, in which the majors reign supreme. It is more advantageous to the producer countries to manipulate supply on the international market in such a way as to arrive at much the same result. For that prupose it is sufficient to limit the production available or to distribute on world markets quantities equivalent to the national holding in the capital of the large companies. This creates a disequilibrium that causes a rise in actual prices. All that then needs to be done is to decide that the posted prices (theoretical prices from which taxation and royalties are calculated) should be brought into line with actual prices. Thus, considerable amounts of additional revenue can be obtained without entering into the international trade in crude, which is a very difficult game for beginners.

The full achievement of that situation has, in fact, been procured by means of the moves made from October 1973 onwards, which constituted a combination of selective and progressive measures for reducing production, of selling oil available from state holdings to the highest bidder (thus hoisting prices to unprecedented levels) and of raising posted prices. The producer countries have now obtained unilateral control over determination of posted prices, which was hitherto the prerogative of the large companies, although that privilege had already been eroded on the occasion of the Teheran agreements in 1971. Furthermore the producer countries have ruled, in line with a decision taken in Teheran in a different context, that posted prices are to be consistently 40 per cent higher than the market prices.

Within a period of one year the posted prices have more than trebled and the actual prices have merely followed or anticipated the movement produced by the measures for curbing production. The rise was accelerated in the second half of 1973; in June a leap of 11·9 per cent, for the purpose of offsetting the effects of devaluation of the dollar, was accepted. On 16 October the Gulf States decided on a 70 per cent rise in posted prices and on 22 December the OPEC countries imposed a further increase, doubling the posted prices. The posted price of oil from the Persian Gulf at the beginning of 1974 was 11·65 dollars per barrel. Libya was not satisfied with that price and since it was unwilling to align itself with the Gulf States dominated by the Shah of Iran, it played the part of a revolutionary by doubling its posted price on 1 January 1974, raising this from 9 dollars per barrel to 18·76 dollars, whilst indicating that it considered the new price to be 'very low'.[8]

In the sphere of prices the policy of the producer countries is assured of success so long as world supplies fall short of needs. There is every reason to regard this as a skilful world leap-frogging exercise. Any decision announced and applied by a group of producer countries is immediately copied or adapted into a more severe action by other producer countries in another region of the world. Thus a rise published in Kuwait or in Teheran leads in time to a chain reaction in Venezuela, in Indonesia, in Algiers, in Tripoli, in Ecuador or in Nigeria, according to the specific quantities of crude involved and the distance from the most important consumer markets. The market has become a generalised world market and, under the pressure of the Arab producer countries, there has been a dissipation of all the partitions which the companies in the 1971 negotiations strove desperately to keep in place. Special preserves, sheltered from the tumult of international affairs, where profitable deals could still be made, have become increasingly rare and rises are reflected throughout the world extremely quickly.

The most important price changes have been mentioned above and the bill which consumer countries will have to pay will be very heavy. But it should be pointed out that before these rises the share received by the producer countries in proportion to the final cost of crude delivered to users had remained small. There are some interesting statistics illustrating this. In 1972 oil exporting countries forwarded 1,300 million metric tons of crude. At the other end of the oil chain users paid for those quantities on receipt 113 thousand million dollars. Out of that total the producer countries collected 17 thousand million dollars, whilst the consumer states obtained 28 thousand million dollars and an equivalent sum went into the funds of the companies.[9] The price rises decided in 1973 will increase the bill by more than 50 thousand million dollars, bringing the revenue of the producer countries to a total of between 25 and 30 thousand million dollars.

It is true that a major change in the distribution of the proceeds is about to be introduced, because of the takeover by the producer countries of the greater part of receipts from oil operations. But there is reason to believe that companies' profits will fall and then rise consistently in line with the income of the producer countries. The current balance sheets of the large companies provide sufficient evidence that the increase in the proportion obtained by the producer countries is not causing a reduction in the profits of such companies. Therefore the governments of consumer countries have a difficult economic choice to make. They cannot pass on the taxes in full to individual consumers, because the energy bill would then be too heavy. It will therefore be difficult to keep indirect taxation at the rates set in the good times of the drop in oil prices before taxation. Those responsible for tax policy will be likely to put tax at a lower level, in order not to penalise the population too greatly. This is a great victory for the producer countries, who will for the first time reverse the overall balance of the taxation of each side. But it is probable that at the same time each country will endeavour in this difficult matter not to act too differently from its neighbours and competitors. An effort to reduce the tax burden will no doubt be made by adjusting the increases according to international circumstances and according to the share of the national market obtained by each petroleum product.

Even if these precautions are adopted, price levels will automatically be lifted by 2–3 per cent by inherent effect of the increases decided at the end of 1973. The world inflationary spiral will be given fresh impetus. This growing inflation will scarcely help towards solution of the oil crisis, because the more world prices climb, the more it is in the interests of producer countries to keep oil in their subsoil when it automatically becomes more valuable as time goes on. Kuwait understood that fact long before the 'October revolution'. It is therefore certain that the figures for production up to 1980 which were estimated before the crisis will not be attained and that the gap between supply and demand will cause a continuing rise in prices in coming years. The only way out of the situation is to expedite European production of hydrocarbons from the North Sea, to take effective measures to prevent wastage and to speed up programmes for making nuclear electricity.

Economic distress

The days of cheap energy are over. The economic system will have difficulty in adjusting to the new strain on it. Branches of industry using huge amounts of energy will be the worst affected. It is known, for example, that the proportion of energy costs in overall costs is 14·7 per cent for glass, 12·2 per cent for dyestuffs, 11·9 per cent for paper and 8·7 per cent for motor car manufacture. Chemical manufacturers producing a wide range of plastics goods will have serious problems of adjustment and industries built around motor transport will suffer from both the rise in production costs and the increased cost of petrol. Moreover, the desire to curb extravagance will mean that private motoring can no longer be favoured to the detriment of the rest of the transport world. All the countries of Europe are in agreement on this point and French planners will have to find industries other than motor manufacturing to assist export efforts. Whilst the prospects for rail transport are good, the future of Concorde, which is a heavy consumer of fuel because of its supersonic speed, is more gloomy than ever, at a time when the U.S. airlines are withdrawing Boeing 747s.

At the stroke of a pen a risk has been created that the balance of payments for each European country will go into deficit or be alarmingly undermined. [10] Britain, already in disequilibrium, will be worse affected than France, which will have to overcome a deficit of more than 2·5 thousand million dollars. West Germany will no doubt achieve a balance that will be only just in equilibrium but for 1974 Japan will have to use the whole of its reserves to meet the higher bill. OECD experts have already tried to work out the extra outflow of currency caused by the successive increases in oil prices. The extra burden was estimated at 15 thousand million dollars a year after the rise of 16 October 1973 and a further 35 thousand million dollars has to be added for the increase imposed on 22 December 1973. This makes a total of 50 thousand million dollars, of which 33 thousand million will be borne by the EEC countries alone. Estimates made at the end of December 1973 for the extra oil expenditures of countries for 1974 in thousand million dollars gave the following list:

United States	9,075
Japan	9,075
West Germany	5,775
France	5,000
Italy	4,125
Britain	4,125
Belgium–Luxembourg	1,650
Netherlands	1,650
Denmark	825
Sweden	1,155
Spain	990
Canada	825
Switzerland	825
Finland	500
Australia	330
Austria	330
Norway	330

These totals need to be revised upwards in order to allow for the cost escalation caused by Libya, which is geographically conveniently placed in relation to Europe, to which it exports the greater part of its production, especially to West Germany. The bill to be paid by Japan is about the same size as that for the United States, and the EEC countries as a whole suffer more than the countries on the other side of the Atlantic. At the same time the American companies are the main suppliers of the Japanese and European oil markets. These companies with headquarters in the United States will have increased profits whilst raising the price of products made by their customers, who are also their competitors in the world market, where commercial rivalry is likely to be intense. Moreover, some American studies put the overall increase for 1974 at 70, not 50, thousand million dollars.

According to White House experts the total cost of world oil imports for the year 1974 was likely to be 115 thousand million dollars, for a volume equivalent to that for 1973 and on the basis of prices at the beginning of 1974, compared with 45 thousand million dollars in 1973.

The producer countries were expected to receive a total of about 100 thousand million dollars.

Details of the expected bill are:

United States: 25 thousand million dollars (compared with 9·3 in 1973);
Western Europe: 55·5 thousand million (compared with 22·2); including 12 thousand million for Western Germany (5·2), 10·7 thousand million for France (3·9), 9·5 thousand million for Britain (3·8), 9·1 thousand million for Italy (3·4);
Japan: 18 thousand million dollars (compared with 6·6);
Developing countries: 10 thousand million dollars.

The oil revenues of the producer countries would increase accordingly: in thousand million dollars Saudi Arabia will receive 19 instead of 3·1, Kuwait 7·8 instead of 1·7, Iraq 5·8 instead of 0·5, Abu Dhabi 3·1 instead of 0·5, Iran 16 instead of 2·5, Libya 6·5 instead of 1·6, Nigeria 6·8 instead of 1·2, Venezuela 10 instead of 2, Indonesia 4 instead of 1.

The figures for the consumer countries are disturbing, even for the United States, which was thought to be less vulnerable. In fact, according to recent calculations, US imports of crude oil (350 million metric tons in 1973) will cost 22 thousand million dollars for 1974, compared with about 9 thousand million for the previous year. This would be equivalent to about 17 per cent of US sales abroad, compared with 6 per cent in 1973.

However, too black a picture should not be painted of the outlook for the industrial countries. For them, the general rise in costs involves highly competitive economies and gives a beneficial spur to industrial ingenuity, in which the Europeans and the Japanese have excelled since the war. Moreover, there will be a better effort to curb extravagance, and the endeavour to improve productivity will oblige industrialists to invest capital in finding means of reducing excessive energy consumption.

The extra amount added to the bill may be regarded as a large transfer of funds from the purchasing countries to the producer states, for which the former will receive some return in one way or another: it is to be hoped that this will not be merely in the form of orders for ultra-sophisticated weapons. There will be lively competition with the United States and the USSR to capture as much as possible of the surplus from the producer countries' receipts in 1974. From the increased revenues at least 20 to 25 thousand million dollars will come back to the OECD area in the form of purchases of goods or services or acquisition of holdings in the enterprises of the industrialised countries.

Such an accelerated flow of funds depends, of course, on the producer countries continuing to place their money according to standards of strict profitability and accordingly not acting too quickly to implement their threat of withdrawing their capital from all Western locations of finance. But a monetary crisis may also be the logical consequence of the present oil war. The capital acquired by the producer countries may aggravate to a fantastic degree the speculation which will arise from the new balance of payments difficulties. There is a danger of the whole system getting out of hand if large movements of wandering capital are involved in a new kind of speculation on even more unsteady rates of exchange.

In fairness, the oil exporting countries ought to assist in the development needs of the Third World, since the producer countries have just imposed an intolerable burden on poor countries which were already fairly heavily in debt and had been preparing to go ahead faster with industrialisation during the next ten years. Whilst in a few cases, such as those of Algeria, Nigeria and Indonesia, there are large deposits of oil in heavily populated countries, the aggregate population of the oil exporting countries in the Third World is only 4 per cent of the total. 67 per cent of the inhabitants of under-developed areas are in countries with a Gross National Product per head lower than 200 dollars a year. For India alone the rise in oil prices will have catastrophic effects on the Five Year Plan beginning in 1974. The oil imports were some 18 million metric tons in 1973 and their cost was equivalent to 10 per cent of the total value of the country's exports.

According to the calculations of *The Petroleum Economist,* India would have to pay 1,241 million dollars for its oil imports in 1974, which is equivalent to 40 per cent of its potential receipts from exports and twice the amount of its foreign currency reserves. For Pakistan, another important sufferer overlooked in the 'October revolution', the bill payable in 1974 for imported oil would be 266 million dollars, which would be more than its foreign exchange reserves for one year. The list of the countries damaged by this oil disaster can be extended, as table 5 shows.

According to a World Bank study, the rise in oil prices will mean an extra cost to the developing countries (some sixty nations) of about 10 thousand million dollars. This will therefore absorb the whole of the external aid granted to them by the OECD countries, which supplies approximately the same sum. The Third World will be out of favour with public opinion in the West and there is a danger of development aid becoming a straightforward subsidy to national funds. After the recent events poor countries striving to develop their economies will suffer two sets of heavy blows, the first from the oil producing countries inflicting prices at three or four times the former level and the second from industrialised countries, whose prices for goods and services will be greatly increased. On top of all this there will be the effects of world inflation at the rate of about 10 per cent a year, which is completely outside the control of the developing countries.

To aggravate this unhappy situation, the overall cost of food for 1973/74 will be about four times that for the previous financial period, at 5·1 thousand million dollars instead of 1·3 thousand million. There is also the rise of at least 500 million dollars in the amount to be spent on fertilisers. The addition to the bill for the poorer nations for 1974 would therefore be expected to be more than 15 thousand million dollars. For lack of adequate currency reserves those countries would have to make drastic cuts in their development programmes, either by reducing oil purchases or by lowering food imports or by restricting the use of fertilisers. The check in progress thus caused would aggravate a bad situation, by increasing the amount of unemployment or by extending the areas subject to famine. The solution could only be provided by action involving both the industrialised countries and the oil-producing countries. The latter would have to grant the poor countries loans in some way guaranteed by the rich countries. The underprivileged nations would then be able to buy not only oil but also the other raw materials and equipment indispensable to them. Is the international conscience strong enough at the present time to rise to this level at which moral obligations and political requirements meet?

The producer countries, these newly rich, containing only a tiny proportion of the world population, are being courted by the other nations. This is a clear sign that they have won a great victory. By cleverly handling the oil weapon they have emancipated themselves from political and economic domination. They now run the risk of attracting rancour

Table 5

Selected developing countries: impact of high crude oil costs (million dollars)

		Oil[a] Estimated cost	Foreign trade[b] Exports	Imports	Balance	Foreign Exchange reserves[c]
India	1972	–	2,372	2,598	(–226)	661
	1973	–	2,934	2,771	163	629
	1974	1,241	–	–	–	–
Pakistan	1972	–	784	644	140	101
	1973	–	983	928	55	254
	1974	266	–	–	–	–
Philippines	1972	–	1,009	1,354	(–345)	309
	1973	–	1,494	1,243	251	606
	1974	693	–	–	–	–
Thailand	1972	–	1,194	1,411	(–317)	838
	1973	–	1,382	1,662	(–280)	1,107
	1974	657	–	–	–	–
Tanzania	1972	–	307	390	(–83)	nd
	1973	–	399	363	36	nd
	1974	62	–	–	–	–
Sierra Leone	1972	–	152	125	27	35
	1973	–	145	125	20	36
	1974	29	–	–	–	–
Sudan	1972	–	355	356	(–1)	38
	1973	–	401	376	25	28
	1974	127	–	–	–	–
Ethiopia	1972	–	183	203	(–20)	53
	1973	–	285	180	105	114
	1974	51	–	–	–	–

[a] Assuming normal requirements and an average landed cost of $10 per barrel

[b] Annual rates based on statistics for the first quarter

[c] At the end of the first quarter of 1973

Quoted in *The Petroleum Economist,* February 1974, p. 47

Source: International Monetary Fund, except 'oil' estimates

and aggression and cutting themselves off from former allies. Already the Peruvian newspaper *La Prensa* has deplored the fact that a handful of countries are dictating to the world, complaining in an editorial of 'economic aggression' in the present attitude of Arab oil-producing countries. The beginnings of solidarity are appearing in Africa, emerging from Algiers, which aspires to retaining the prestige of being the capital of the non-aligned countries and has always preached a doctrine that hydrocarbons were not an end in themselves but had the function of giving impetus to the overall development of a country. This new solidarity seems particularly urgent, as the Sudanese Minister of Foreign Affairs, Mr Mansour Khalid, has stated that the countries of Africa may be expected to spend one thousand million dollars on oil imports in 1974, compared with 400 million dollars in 1973. Mr Khalid, who is chairman of the special commission of the Organisation for African Unity set up to study the consequences of the oil crisis for the economies of the African countries, commented that 'the oil-producing countries did not consult the other members of OAU before taking their decisions'. He declared that:

> Ways must be found of helping the innocent victims, through no fault of their own, of the oil crisis, such as the African countries. The producer countries should ensure normal delivery of oil to the African countries and grant them the benefit, in one way or another, of a preferential price system.
>
> If that is not possible, then a scheme of long-term credits or of payment facilities must be considered. It may be difficult for the producers to have two price systems for their oil but there are other alternatives and ways can be found.

Despite the efforts of Algeria and Iran, it seems unlikely that this cry for help will be heeded. In that case sufferers might become increasingly aggressive towards the countries which obtain profit from their resources of black gold without labouring for their gains. If crude oil prices are constantly pushed up the Arab countries will be held responsible for world inflation and for the social and political troubles which in time result from it. Public opinion in the consumer countries may lose patience and support retaliatory measures by their exasperated governments. It is necessary to realise the serious implications of the threat of an energy war, which would destroy the unstable international detente and would shatter the present Soviet-American co-operation in watching for and acting promptly to suppress any dangerous disturbance anywhere in the world.

The new policy of the producer countries will once again strengthen the most powerful, namely countries potentially self-sufficient in oil, principally the United States, the USSR and China. The United States will put its full weight into an endeavour to draw Europe into whatever Atlantic schemes suit the objectives at any given moment, whereas the interests of Europe lie in devising new forms of co-operation with the countries of the Mediterranean basin and of the Near and Middle East.

The triumph for oil undoubtedly marks the beginning of its decline. The oil producing countries which have no other natural resources have 10 to 15 years left in which to lay the foundations, by means of the application of the capital received, of development that will eventually be able to manage without this non-renewable product. They will need to continue to finance exploration for hydrocarbons and to find new deposits in order to replace those now being used up, because the foreign companies will gradually pull out from producer countries where they obtain little benefit from oil production and the host countries are too greedy for the rewards. The foreign companies will henceforth look for profits in downstream operations, which would appear to bring their interests into greater accord with those of the consumer countries. It would be worthwhile to take a partnership holding in the overall enterprise in these circumstances, as Iran has realised, because the companies will have to give ground in production and spread their interests over refining, petrochemicals and distribution in the territories where they have hitherto been dominant. But they will also need to try to move into alternative lines of production (coal, shale oil and hydrogen) and into the nuclear energy field, in order to find new bases of power away from the areas of operations where governments are trying to deprive them of influence.

The financial measures taken by governments are moving in that direction. Since security, the quest for minimum prices and the desire for autonomy are the predominant considerations in any energy policy, the bulk of the public and private capital available in the consumer countries will be directed towards areas where Western technology may operate freely, without excessive dependence.

Certainly Europe and Japan will experience 15 difficult years, because it takes time to overcome problems of such magnitude. According to present estimates the share of nuclear energy in electricity production in Europe will be 21 per cent by 1980 and 32 per cent by 1985. But by 1985 the North Sea will be providing large quantities of oil, because investments there can be multiplied without appearing to be excessive in the prevailing circumstances. By that date Britain will have become self-sufficient in oil and Norway will be exporting crude. New sites of oil deposits will have been found. It was recently reported that only 5 per cent of the sedimentary basins belonging to Brazil had been tested by geophysicists. Any price increase renders exploration less costly, especially the search under the sea which is involved in 50 per cent of the concessions held by the large oil companies of the West. There are new, less politically troubled areas awaiting the adventurers of this era, in the Atlantic, the Antarctic and the Mediterranean.

In the present crisis oil has become a political weapon, inflicting damage on the industrialised countries but still more so on the poor countries. It will have to be reckoned with over the next ten years. But its triumph probably also marks the beginning of its irremediable decline. Sometimes weapons inflict damage on both parties in the struggle. And it has been known since methods of warfare were first invented that one kind of weapon is especially dangerous to the user – the boomerang.

Notes

[1] Quoted in *Le Monde*, 9–10 December 1973.

[2] Quoted in *Pékin Information*, no. 48, 3 December 1973.

[3] *L'industrie française du pétrole*, 1972, p. 2.

[4] Ibid., p. 3.

[5] Saudi Arabia, Kuwait, Libya, Iraq, Algeria, Abu Dhabi, Egypt, Qatar, Syria, Bahrein.

[6] Comité Professionnel du Pétrole, *Pétrole 1972*, statistical material.

[7] *The Petroleum Economist*, May 1974, p. 175.

[8] Yet the price for Libya had been fixed too high in relation to the other producing regions. It had to be discreetly brought down to a fixed price at a lower level. Moreover, the OPEC countries are not unanimous on this matter. It is well known that if the Arab countries are all in agreement on keeping the level of prices obtained immediately after the crisis, then Saudi Arabia will press for devaluation, with the active support of the United States. Furthermore, when rises occur in such rapid succession the posted price becomes completely out of touch with the levels found in actual transactions. At the end of the Geneva Conference (7–9 January 1974) the producer countries decided that henceforth taxation was to be based on actual prices and the ratio 1·4:1 between posted prices and actual prices was to be abandoned.

[9] These figures are from *Le Monde*, 25 December 1973.

[10] See OECD, *Perspectives économiques*, December 1973, the chapter entitled 'La crise pétrolière'.

Conclusion: End of an Era

We are witnessing the end of an era. A new age in oil affairs has just begun. The bitterness of the confrontations between the various participants serves to show up more clearly the relative stability of the last decade.

To conclude this book the findings and perspectives obtained in the analysis that has been made will be summarised briefly. Then, after a review of the recent history of oil affairs, the events of the 1970s will be discussed and the outlook for the future will be considered. The years 1970–73 mark both a point of arrival and a new point of departure. In connection with oil both continuity and change may be noted. At the moment oil still belongs to the economic world but it is increasingly becoming a political issue.

Findings and perspectives

The respective aims of the consumer countries and the producer countries on either side of the Western Mediterranean have been examined. Then the possibility of co-operation – however slight – between relatively contradictory aims was investigated. It was pointed out that despite their opposing positions the countries supplying oil and the countries using it still had common interests. The concept of 'rival associates' seemed to be a means of assessing this paradoxical situation: whilst striving on behalf of its own attitude – and thereby enhancing antagonism between states – no country in the region under consideration can afford to break off relations with its opposite numbers completely; because of implicit or explicit alliance none of the participants in oil affairs really has the means of winning a definitive victory over the others, unless it is prepared to withdraw from the contest altogether, deliberately scrapping the rules of the game which all those concerned regard as essential.

The extent to which governments intervene in oil operations has been studied. In examining the relationship between avowed intentions and the means of achieving them, two types of government action were considered, namely direct management and indirect interventionism. This led to noting the merits and some of the weaknesses of the state companies. It was also pointed out that although interventionism is amply justified, there are long-term risks which are beginning to be perceived. Governments are endeavouring to steer a difficult course, which varies according to the circumstances of the era, between uncontrolled instability and unsuccessful inflexibility. For a long time the industrialised consumer countries held the initiative but this has now passed to the producer countries. As a result, the ties which have existed between the various participants in oil affairs in the Western Mediterranean region may in future be weaker. Yet it is difficult to believe that they will disappear altogether if, in the nature of things, the countries with Mediterranean coastlines are destined to remain 'associates', even if they become more strongly 'rivals'. However, another possibility is that the predominance of oil as a form of energy has already reached its zenith, since oil curves are already rising less steeply and nuclear energy is likely to make increasing progress.

The state of oil affairs

At the beginning of this book the situation in Europe in the second half of the 1950s was chosen as the starting point for the study. Up to that time Europe had experienced a shortage of oil, in a kind of special extension of the general scarcity of materials after the war. But from 1957–58 onwards competition, which had been feeble, began to be fiercer. Following the discoveries of oil in the Sahara and in Libya, so close to industrialised Europe, oil supplies became abundant. The market changed into a buyers' market, and thus became advantageous to the EEC area, which was dedicated to the principle of lowering internal customs barriers.

Now the situation is reversed, with a sellers' market enabling the producer countries, banded together in OPEC, to seize considerable advantages, to the detriment of consumers in the industrialised countries. The three main participants in oil affairs (producer countries, consumer countries and oil companies) are still involved in an interplay of relationships but their positions have been altered. Oil continues to create conflicts where political and economic interests become entangled with each other but there is a complete shift of emphasis. Important changes are taking place, even though the structures from the previous age may appear to remain, and will in time make the circumstances of encounter between the parties quite different.

In the light of those general points some conclusions are now presented. These will be arranged under four time periods, namely oil affairs of the

1960s, the turning point in 1970–71, the crisis of October 1973, and possible future developments.

A decade in oil affairs, the 1960s

During the 1960s the consumer countries tended to be at a slightly greater advantage than the other two parties in relationships between companies, producer states and consumer countries. True, the producer countries achieved some of their fiscal objectives but their alliance in OPEC was late in affecting the countries of North Africa and so far moderate views had prevailed over aggressive attitudes. The decline of posted prices was arrested but there was no marked upturn. The abundant supplies of crude available after expansion of production in Algeria and Libya chiefly benefited consumers, who managed to obtain substantial reductions in actual prices, as a result of competition between oil companies. The West European countries with Mediterranean coasts managed to obtain petroleum products cheaply enough for oil to rise over a ten year period to a lead over other sources of energy, to the detriment of coal, which had hitherto hoped to keep ahead of oil. Another important aim of the consumer countries, security of supplies, was fairly well satisfied. The 1967 crisis, following the second Arab-Israeli war, proved to be less damaging than the previous emergency. Lessons had been learnt from the 1956 upheaval. Moreover, during that decade the consumer countries did not allow themselves to become mere buyers. They were able to reduce their dependence by bold commercial policy and by entering into the intermediate stages of oil operations (transport, refining and petrochemicals).

This degree of autonomy was in part obtained by setting up state companies. At one point it was possible to wonder whether those companies, being bolder, more closely connected with political alliances and regarded with less suspicion by the producer countries in connection with 'colonialism', might adversely affect the traditional positions of the large international companies. However, this idea underestimated the powers of resistance of the majors. Any decline suffered by the latter was largely to the benefit of the 'independents', which were for the most part American enterprises, rather than in favour of the state companies of the consumer countries. Through their newly-acquired strength, mainly from operations in Libya, the leading independents to some extent profited at the expense of the less well placed of the majors. None of the state companies was in such a fortunate position. Circumstances during the 1960s greatly assisted the endeavours of the public corporations, but not to the point of allowing them to displace the firmly established majors. Favourable circumstances may give newcomers a chance but they cannot produce a great change in long-standing arrangements. That fact was apparent at the conferences in Teheran and in Tripoli at the beginning of 1971, at which the oil enterprises needed the assistance of the majors and chose them as spokesmen. ENI's somewhat hypocritical protestations do not alter the realities of the situation. At the moment of truth everyone was obliged to take sides. The dividing line passed between producer countries and oil companies of the industrialised countries. From that point of view the years 1970 and 1971 definitely marked the end of one era and the beginning of another.

Turning point of the years 1970–71

The years 1970 and 1971 constitute a period of change when the situation was completely re-examined. Up to the end of 1969 Europe had for a number of years experienced a kind of euphoria where oil was concerned. Oil had been arriving regularly at the ports. Sellers were giving rebates in a fight with each other for customers and in order to find buyers for increasing quantities of petroleum products. Coalmining in France, and later in Germany, was unable to hold its own against such strong pressure from a competitor at such an attractive price. Little attention was paid to those who advocated the maintenance of costly security based on solid fuel. The newly-built thermal power stations used fuel offered under long-term contract on conditions defying competition from any other energy source. Programmes for nuclear power stations were postponed because they did not afford any prospect of competing satisfactorily.

The movement of a few grains of sand sufficed to cause distortion of this fine edifice. At the end of 1969 the European energy market underwent a complete change.

The first signs of upheaval appeared at Rotterdam, the oil port where prices for additional tonnages needed by refiners operating on a large scale are decided. Heavy fuel oil was still selling at 9 dollars a ton in September 1969 but rose to 24 dollars by the end of 1970.

There were many reasons for this price rise. There was an insatiable demand from Japanese industry for energy sources. Because of this the Japanese interests were taking every opportunity to buy in all the main markets in the world, which put the price up. The efficacy of the fight against pollution in the United States obliged middlemen to give up obtaining supplies from 'dirty' coalmines and to buy in Europe the North African petroleum products with lower sulphur content than those on the domestic market in the USA. Consumption was increasing so fast, having doubled in nine years, that it was outdistancing the estimates of the experts, especially in cold weather – because of the rise in consumption of domestic oil. Supply from the Mediterranean region had fallen short of the companies' expectations. Libya had unilaterally imposed restrictions of production in 1970. Tapline, carrying Saudi Arabian crude to the Mediterranean, had been closed for nine months, because of an 'accident' to the pipeline in Syrian territory.

The gap between supply and demand caused enterprises to turn more definitely to the oilfields in the Middle East, but there was insufficient tanker capacity available to cope with this sudden rush of trade. Hence there was a normal rise in oil prices and an abnormal speculative rise in freight rates from the Persian Gulf to Europe.

Because of this world situation the advantage suddenly tipped in favour of the producer countries banded together in OPEC. They could not miss such an obvious opportunity. After having staked everything on oil over the last ten years, Southern Europe had to give way on its two predominant objectives of seeking security and of obtaining energy at reasonable cost. The producer countries knew how to play on anxiety over those two aims in order to achieve what they themselves sought. They took up a firm stance midway between these two objectives. By threatening to cut off supplies when Europe was worried about its insecurity they managed to achieve acceptance of large rises in their rates. By applying a certain amount of blackmail in connection with security of supplies they made the consumer countries accept the delivery of expensive goods. The companies acted as go-betweens and only managed to survive by passing on the extra costs almost in full to consumers.

Crisis of autumn 1973

Consumers were the main victims of the crisis in the autumn of 1973, which has already been analysed. Europe is again afraid of shortage and has again become anxious about security.

'Europe would have frozen to death this winter if settlement had not been reached in the Middle East', said President Nixon at a press conference. Others think it more appropriate to refer to 'an oil Munich', in view of the way in which European and Japanese negotiators responded to the highest demands of the Arab producer countries.

An earthquake has shaken industrial economies, which are heavy consumers of energy in the form of oil. The effects began to be passed on within a few months, but the full extent of the consequences was not immediately envisaged. There was much talk about petrol but it only represents a small proportion of the total consumption of petroleum products in Europe. On looking at the 1972 figures for France one finds that petrol constituted about 15 per cent of overall consumption of petroleum products, whilst domestic fuel oil and heavy fuel oil for industrial use accounted for 34 per cent and 30 per cent respectively.

The Middle East producer countries, holding 60 per cent of the present proved oil reserves in the world, undoubtedly have the power to curtail the movement of motor vehicles, to cause heating and lighting of offices and dwellings to be restricted and to put employees of industries with heavy fuel consumption out of work.

But matters are not likely to be as bad as that. Indeed, if we were threatened with such a disaster there would be talk of war and there would be a spirit of war in the populations of the wealthy countries if they were deprived of jobs, comfort and means of travel.

There has certainly been a victory by the producer countries. They have gained control of quantities and of prices, and by manipulating these two items they can quietly obtain the equivalent of nationalisation. The functioning of the industrial economy will depend on the Arabs and for the next ten years there is no credible alternative to submitting to their pressure. At present there is no adequate substitute for 'black gold'. Income will pour into those countries and the money of the users will fill the coffers of the oil producing nations.

However, it is an uncertain and limited victory. Worst affected will be the countries of the Third World, now painfully striving to achieve industrialisation. Having no resources in their own territory, India and Pakistan will also have to pay the much higher prices for imported oil. Are there any countries in the Third World that could use a weapon comparable to the one which has been taken up by the oil-producing countries? No doubt the countries producing copper or certain rare minerals are in a similar position. For the other countries, the majority, more than two-thirds of their national exports comprise agricultural produce and the rich countries tend to protect themselves against food shortages. Can one imagine India or Cuba threatening to stop deliveries of textiles or sugar in order to obtain better prices? A suitable strategy for redressing the deterioration of trading conditions is difficult to find in the case of oil. Moreover, the higher energy prices become, the greater the risk is that manufactured goods exported by the rich countries to the poor countries will in turn be more expensive.

By a geographical accident oil is exported in abundance by countries that are among the lowest populated in the Third World. Whilst achieving one-sixth of the value of world exports from the least developed regions, in 1969 the oil producing countries of the Middle East had only 3 per cent of the population of those regions, or 1·5 per cent if Iran is excluded. True, the rise in the prices of hydrocarbons will make progress easier to attain in Algeria, in Nigeria and in Indonesia. Iraq, Iran and Venezuela will be able to begin a real process of development. But what about the others? There is no time to lose. Soon new sources of energy, for example atomic energy, will be available, or the older ones, including coal, will be rendered more economic by excessive increase in oil prices. Then the last vital link between rich countries and poor nations will be severed.

Supplies of textiles and natural rubber, which provided a foundation for the industrial revolution in Europe and in America, or cheap food, which formed the basis for the international distribution of labour advocated by Ricardo, are no longer the 'matter of life or death to all civilised nations' to which Marx and Engels refer in the Communist Manifesto of 1848.

The 1973 crisis is likely to give tremendous impetus to exploration for other oil deposits and to the development of new sources of energy. Investments will not go to the unstable oil regions but will be employed in more remunerative sectors which are less politically risky. The rise in prices will hit Japan, Europe and the United States hardest in the industrialised world. The energy bill has always been high in the US, which has rightly been described as a paradise for small and medium-sized oil enterprises.

Above all, the large American companies, which extract oil in the East and sell it to European and Japanese users, will benefit from the situation. At the same time as they increase their profits they put up the prices of the goods made by the enterprises which buy the oil, and the latter are competitors of the United States in the world market. Saudi Arabia is destined to play a prominent part in the future development of oil production and will receive increased revenue. Because of that country's very close links with its American partners, a large amount of the money collected will go back to US banks, thus creating the paradoxical situation that the true origin of part of the US capital resources will be European and Japanese purchases.

However, the picture is not wholly gloomy. Hardship may impel the countries of Europe to close ranks and put an end to the anomaly of great economic potential but very little political power. Wastefulness of economic resources will be less and less tolerable and the obsession with annual growth rates will be abated. The emphasis may shift from standards of living to ways of living. The motor car's reign will be challenged. When the means of living come under some restraint, then it will be easier to question the reasons for ways of living. The oil crisis will prove to be a blessing if it obliges the Europeans to discover new techniques for hastening the building of a society in which life is really pleasant.

Uncertainties regarding the immediate future

Thus, there are some question marks about the immediate future.

Energy production within the national territory of countries of Southern Europe will continue to decrease. According to the Energy Commission report on France's Sixth Plan, domestic energy production in 1960 corresponded to 62 per cent of France's needs, and at the end of the Fifth Plan it amounted to only 39 per cent; in 1980 it will be not more than 20 per cent. The share of the Middle East and North Africa in supplying France's energy was 24 per cent in 1960, had risen to 48 per cent in 1970 and will be about 60 per cent in 1975. This increasing dependence is likely to be very costly. Official limitation of the annual total of amounts imported may enable the process to be checked but cannot cause a decrease between now and 1980.

Means of stopping this outflow of expenditure cannot be found other than by augmenting the Merchant Fleet, increasing the tonnages of underground stocks, retaining the possibility of using coal for the thermal power stations which have changed to fuel oil, boosting oil production by national enterprises and trying to obtain a positive balance in trade in refined products.

With spectacular increases in the amount of energy supplied by foreign producers there is a risk of an inflationary process being created almost automatically. The branches of industry which use the most energy would be worst affected.

However, it should be noted that the additional revenue paid to the producer countries will result in increased purchases of durable goods from the industrialised countries: thus they will be the first to benefit from the rises imposed by their suppliers. Thus we come back to a recurrent theme in this book, namely that in order to grasp the meaning of the energy bills presented to importers it is necessary to think in terms of overall balance – of the flow of funds into and out of the country. The rich countries of the West will be in a more fortunate position than the countries of the Third World which lack oil of their own and are unable to supply goods and services of equivalent value in order to balance more expensive purchases. The victory of the producer countries is a victory of a small number of poor countries. The bill will be heavy for poor countries which lack oil riches in their subsoil and which are engaged in a costly industrialisation process.

On the other hand the United States and, probably, the countries of the Communist bloc will reap real benefits from this rise in the cost of energy supplies to Western Europe. Since the price of oil in Europe has almost caught up with the protected price in the US, the Americans will gain in two spheres. They will be able to bring back into production the oilwells which in recent years have been abandoned as uneconomic. At the same time, through a few large companies with worldwide scope of activities they will be able to go on supplying Japanese and European consumers with crudes so expensive as to reduce the competitivity of the national industries and put American enterprises in a favourable position.

However, it is uncertain how long the majors will be able to persist with the contradictory policies of obtaining good profits and currency earnings in the countries where they have bases, yet presenting themselves as the sole protectors of consumers and customers.

The Algerian decisions on partial nationalisation of the French oil companies have for the time being tended to strengthen the majors, because the initial result has been to destroy a spirit of co-operation between producers and consumers which was beneficial to both parties. ERAP's strategy has failed in Algeria not because it was wrong but because the organisation has not had either the time or the means to expand elsewhere. But the most progressive oil-producing countries have realised in time that possession, under their sovereignty, of their national resources is as important to national development as profits from prices. Accordingly the 1973 oil crisis has drawn attention to the quantitative and qualitative aspects of co-operation between industrialised countries and countries producing primary materials. True co-operation can only occur between sovereign nations which are in control of their own fate and are free to choose their schemes for development. If that concept has been accepted, even if it is put into practice by force, as in the case of Algeria – and according to surveys the majority of the people of France concur with it – this surely signifies an important change.

Now that Algeria, as an oil-producing country, has its own oil production company it is trying to promote that view to other exporting countries – and the example may in time prove to be contagious. The Gulf States have obtained holdings in foreign companies operating in their respective territories. The country which is staking its claim to holdings most strongly is Saudi Arabia, which has hitherto been a vassal of the American companies and has justifiably been considered the opposite of a revolutionary country.

The time is coming when it will be necessary to invent new forms of co-operation enabling the producer countries, guided by determined men and desiring change, to nationalise foreign assets in whole or in part. The industrialised consumer countries possess skilled men, technological resources and finance and they should be able to find better forms of co-operation with these newly 'sovereign' states. What will be required is assistance to the possessors of the raw material to enable them to change from taxing the proceeds of oil operations by other people to treating oil as a sector of the domestic economy. The rules of the oil game are such that none of the players can achieve definitive victory over the players on the opposing sides. However the rivals can always become associates. Political debate should exercise supremacy over economic confrontation and not the reverse.

In practice it would be advisable to seek forms of co-operation in the nature of agency or entrepreneurial agreements. Some such agreements were introduced timidly in 1966 in the Middle East. The new formula needed may be envisaged as an association having the essential characteristic that the foreign participant acts as a mere entrepreneur; it is merely a provider of financial, commercial and technical services. The foreign concern performs, in fact, the combined rôles of banker, supplier of advanced technology and broker. As a supplier of funds its function is to lend the money required to put into operation the working programme which the foreign concern and the national company of the host country have jointly designed. There can be no repayment of sums advanced for exploration until there has been a discovery sufficiently large for commercial viability and repayment must be according to conditions favourable to the national company, such as no interest or redemption of the loan over a long period (fifteen years, for example). Loans for development would be repayable more quickly and at a moderate rate. If so requested by the national company, the Western contractor may undertake to place the crude on foreign markets and take a prescribed commission. But clearly the foreign concern remains a technical assistant and mere operator. It puts its capabilities of all kinds at the service of its associate. The latter remains the owner of the installations and also of the crude discovered and extracted. All the phases of development are theoretically performed under its sole control. The co-operating concern is rewarded as an enterprise rendering services, i.e. it is entitled to purchase a given percentage of production at a price slightly above cost price.

Such is the general design and content of agreements which are likely to be required in future as a means of changing over from the excessive dependence in which many countries with oil in their subsoil still find themselves. Unless such new formulas are devised it is difficult to see how the finance indispensable for increasingly extensive and costly exploration can be supplied or how the wastage caused by the unhealthy effects of a nationalistic attitude can be avoided. Action along these lines would also be a positive measure for achieving collaboration rather than increasingly violent clashes between the rich countries, which are becoming better and better equipped in human, financial and technological resources, and the poor countries, which are shunned and treated with rancour and are so short of means of acting effectively in the context of world competition.

As I said at the beginning of this book, oil is a commodity around which relations of co-operation and of rivalry between men, institutions and states form. In attempting to unravel the complicated web of the many interests involved in the region of the Western Mediterranean chosen for special study my intention has been to emphasise that oil, which is located at a crossroads of economics and politics, is increasingly becoming a political issue.

But obviously politics reshape the field of the various interests in a way that the economist does not always find coherent. The wish to maintain a certain national consensus, the needs of international relations, the degree of ability to foresee events, the evaluation of possible risks and the lessons of past failures are all factors which undoubtedly cause a politician to promote certain aspects and neglect others. As B. de Jouvenel commented in his book *De la politique pure,* political problems have the special feature that there is no solution to them. They are resolved by compromises, by settlements or by temporarily coming to terms. But they are not solved by solutions in the sense that a problem of geometry has a solution. All things being equal, an economist is capable of giving one or more solutions to a specific problem. Strictly speaking politics cannot: there is a political problem precisely because there is no solution. This paradox is nowhere more apparent than in the development of the politics applied to oil economics in the countries with Western Mediterranean coastlines.

Bibliography

Oil as an economic matter

Berreby, J.-J., *Histoire Mondiale du Pétrole,* Ed. du Pont Royal, Paris 1961.

Bradley, P.G., *The Economics of Crude Petroleum Production,* North Holland Publishing Company, Amsterdam 1967.

Campbell, R.W., *The Economics of Soviet Oil and Gas,* J. Hopkins Press, Baltimore 1968.

Cattan, H., *The Evolution of Oil Concessions in the Middle East and North Africa,* Dobbs Ferry, New York 1967.

Chapelle, J., *Economie du Pétrole. Le Pétrole dans le Monde,* Technip, Paris 1964.

Chardonnet, J., *Géographie Industrielle,* vols 1 and 2, Sirey, Paris 1962–65.

Chevalier, J.-M., *Le Nouvel Enjeu Pétrolier,* Calmann-Levy, Paris 1973.

Flandrin, J., and Chapelle, J., *Le Pétrole,* Technip, Paris 1961.

Frankel, P.H., *L'économie Pétrolière: Structure d'une Industrie,* Librairie de Médicis, Paris 1948.

Grenon, M., *Ce Monde Affamé d'Energie,* Preface by Sicco Mansholt, Ed. Robert Laffont, Paris 1973.

Longhurst, H., *Adventure In Oil,* William Clowes and Sons, London 1959.

Mainguy, Y., *L'économie de l'Energie,* Dunod, Paris 1967.

Majorelle, J., *L'économie de l'Energie,* IEP Course, Paris 1966–67.

Masseron, J., *L'économie des Hydrocarbures,* Technip, Paris 1969.

Penrose, Edith, 'International economic relations and the large international firm', pp. 107–36 in: Edith Penrose (ed.), *New Orientations. Essays in International Relations,* Frank Cass and Company Limited, London 1970.

Puiseux, L., *L'énergie et le Désarroi Post-Industriel, Dossier Méthodologique de l'Energie et le Désarroi Post-Industriel,* Hachette, Paris.

Tugendhat, C., *Oil, the Biggest Business,* Eyre and Spottiswood, London 1968.

Tugendhat, C., *Freedom for Fuel,* Institute of Economic Affairs, London 1963.

Zimmermann, E.W., *Conservation in the Production of Petroleum,* Yale University Press, New Haven 1957.

Oil prices

Cassady, R., Jr, *Price Making and Price Behaviour in the Petroleum Industry,* Yale University Press, New Haven 1954.
Frank, H.J., *Crude Oil Prices in the Middle East. A Study in Oligopolistic Behaviour,* F.A. Praeger, New York 1966.
Laudrain, M., *Le Prix du Pétrole Brut, Structures d'un Marché,* Librairie de Médicis, Paris 1958.

Oil as a political instrument

Aron, R., *Sur la Politique Pétrolière,* unpublished.
Bauchard, D., *Le Jeu Mondial des Pétroliers,* Le Seuil, Paris 1970, (Coll. Société).
Baumier, J., *Les Maîtres du Pétrole,* Julliard, Paris 1969.
Berreby, J.-J., *Le Pétrole dans la Stratégie Mondiale,* Ed. Casterman, Paris 1974.
Durand, D., *La Politique Pétrolière Internationale,* PUF 3rd edition, Paris 1970, (Que sais-je? No. 891).
El-Sayed, M., *L'OPEP* (OPEC), Librairie générale de droit et de jurisprudence, Paris 1967.
Frankel, P.H., *Oel-Tatsachen und Tabus,* Verlag für Literatur und Zeitgeschehen, Hannover 1963.
Hartshorn, J.E., *Oil Companies and Governments. An Account of the International Oil Industry in its Political Environment,* Faber and Faber, London 1962.
Lenczowski, G., *Oil and State in the Middle East,* Cornell University Press, New York 1960.
Mosley, L., *La Guerre du Pétrole,* Les Presses de la Cité, Paris 1974.
Mughraby, M., *Permanent Sovereignty over Oil Resources,* Middle East Research and Publishing Center, Beirut 1966.
Odell, P.R., *Le Pétrole et le Pouvoir Mondial,* Ed. Alain Moreau, Paris 1973.
Odell, P.R., *Oil and World Power. A Geographical Interpretation,* Peter Hall, London 1970, (Pelican Geographic Series).
Odell, P.R., *Oil: The New Commanding Height,* Fabian Society, London 1965.

Revue Française de Science Politique, XII (6), December 1972, special issue on oil conflicts, 1970–71.
Sarkis, N., *Le Pétrole et les Economies Arabes,* Librairie générale de droit et de jurisprudence, Paris 1963.
Shwadran, B., *The Middle East, Oil and the Great Powers,* F.A. Praeger, New York 1959.
Tanzer, M., *The Political Economy of International Oil and the Underdeveloped Countries,* Beacon Press, Boston 1969.

Oil and the French state companies

Chenot, B., *Organisation Economique de l'Etat,* Dalloz, Paris 1951.
Rapport sur les entreprises publiques, dit Rapport Nora, La Documentation française, Paris 1968, (Report).

Oil and Spain

Tamames, R., *Los Monopolios en España,* Editorial Zyx, Madrid 1967.
Tamames, R., *La Lucha Contra los Monopolios,* Technos, Madrid 1966.
Tamames, R., *Estructura Económica de España,* Sociedad de estudios y publicaciones, Madrid 1964.

Oil and France

Clair, P., *L'indépendance Pétrolière de la France,* vol. 1, *Le Théâtre de Guerre,* Ed. Cujas, Paris 1969.
Faure, E., *Le Pétrole dans la Paix et dans la Guerre,* Editions de la Nouvelle Critique, Paris 1939.
Fontaine, P., *L'aventure du Pétrole Français,* Les Sept Couleurs, Paris 1967.
Murat, D., *L'intervention de l'Etat dans le Secteur Pétrolier en France,* Technip, Paris 1969.
Oizon, R., *L'evolution Récente de la Production Energêtique Française,* Librairie Larousse, Paris 1973.
Touret, D., *Le Régime Français d'Importation du Pétrole et la CEE,* Librairie générale de droit et de jurisprudence, Paris 1967.
Vilain, M., *La Politique de l'Energie en France de la Seconde Guerre Mondiale à L'horizon 1985,* Ed. Cujas, Paris 1969.

Oil and Italy

Bruni, L., and Colitti, M., *La Politica Petrolifera Italiana,* A. Giuffre, Rome 1967.

Faleschini, L., and Kojanec, G., *Ente Nazionale Idrocarburi (ENI),* Editore Carlo Colombo, Rome.

Frankel, P.H., *Mattei: Oil and Power Politics,* Faber and Faber, London 1966.

Guarino, G., *L'intervention de l'Etat Italien en Matière d'Hydrocarbures,* in: Annales de la Faculté de droit d'Aix, 52, 1960–61, pp. 41–60.

Mattei, E., *Problèmes de l'Energie et des Hydrocarbures,* ENI, Rome 1961.

Posner, M.V., and Woolf, J.J., *Italian Public Enterprise,* Gerald Duckworth and Co., London 1967.

Votaw, D., *The Six Legged Dog, Mattei and ENI: A Study on Power,* University of California Press, Berkeley 1966.

Oil and Algeria

Amin, S., *L'économie du Maghreb,* 2 vols, Ed. de Minuit, Paris 1966.

Dubreuil, J., *Le Pétrole Arabe dans la Guerre,* Ed. Cujas, Paris 1968.

Gendarme, R., *L'économie de l'Algérie,* A. Colin, Paris 1959, (Cahier de la Fondation Nationale des Sciences Politiques, no. 101).

Mainguy, M., *Le Pétrole et l'Algérie,* Le Cerf, Paris 1958.

Tiano, A., *Le Maghreb Entre les Mythes,* PUF, Paris 1968.

Tiano, A., *Le Développement Economique du Maghreb,* PUF, Paris 1968, (Collection SUP).

Main periodicals consulted

Petroleum Press Service, PPS
Petroleum Intelligence Weekly, PIW
Bulletin de l'Industrie Pétrolière, BIP
Middle East Economic Survey, MEES
World Oil
Journal of International Affairs
The Economist
Orient Arabe
Pétrole et Gaz Naturel Arabes
Revue Française de l'Energie, RFE
Revue Algérienne des Sciences Juridiques, Economiques et Politiques

Revue Française d'Etudes Politiques Africaines
Marchés Tropicaux
Coopération Technique
Humanisme et Entreprise
Direction
Les Echos
Expansion
Entreprise
Projet
Revue de Défense Nationale
Revue Française de Science Politique
Stratégie
Bulletin Mensuel d'Informations ELF, BMI
ENI Informazioni
CEP, TOTAL, Informations
ELF, Report of the Management (annual)
ENI, Exercise Annuel, Summary of Board's report (annual)
ENI, Relazioni e Bilancio, as at 31 December each year
La Razette, house journal of *Sonatrach*

Sundry publications consulted

International Petroleum Encyclopedia, Tulsa, United States.
Chase Manhattan Bank, *Capital Investments of the World Petroleum Industry* (annual).
Chase Manhattan Bank, *Financial Analysis of a Group of Petroleum Companies* (annual).
UNO, OECD and EEC publications.
Annuaire de l'Afrique du Nord, Aix-en-Provence, Editions du CNRS (annual).
Publications of Comité professionnel du pétrole, Paris.
Publications of IFP, 'Colloques et séminaires' collection, Paris.
French Ministry of Industry, DICA, *Activité de l'Industrie Pétrolière,* Paris (annual).
Commissariat général du Plan, Commission de l'Energie, Comité du Pétrole, Paris.
Bulletin de l'Economie et des Finances, Paris.
Notes et Etudes Documentaires, Documentation Française, Paris.
International Labour Organisation, Oil Commission, 8th session, *Industrial Activities Programme,* Geneva 1973.

Index

Indexer's note: Certain references may lead only to footnote indices; if the subject cannot be found on the page indicated, the reader should consult the footnotes at the end of the chapter which will give him the necessary lead.